D1713491

PROBLEM SOLVING THROUGH SYSTEMS THINKING

HOW TO APPLY SYSTEMS THINKING TO TACKLE PRESSING ISSUES AND UNFOLD A CLEAR SOLUTION IN A HIGHLY INTERCONNECTED WORLD

WISDOM UNIVERSITY

CONTENTS

For Our Readers 1

What Reader's Are Saying About Wisdom University 9

Introduction 13

1. What Exactly Is A System? 19
2. (Extra)Ordinary Encounters With Systems 37
3. Now, Let's Talk About Systems Thinking 55
4. System Secrets 71
5. Navigating The Complexities Of The Big Picture 99
6. Beneath The Systems Blueprint 121
7. Break The Mold 143
8. How To Dance With Systems 164

Afterword 201

Over 10,000 People Have Already Subscribed. Did You Take Your Chance Yet? 203

The People Behind Wisdom University 205

References 209

Disclaimer 223

Get ALL our upcoming eBooks for FREE

(Yes, you've read that right)

Total Value: $199.80*

You'll get exclusive access to our books before they hit the online shelves and enjoy them for free.

Additionally, you'll receive the following bonuses:

Bonus Nr. 1

Our Bestseller

How To Train Your Thinking

Total Value: $9.99

If you're ready to take maximum control of your finances and career, then keep reading...

Here's just a fraction of what you'll discover inside:

- Why hard work has almost nothing to do with making money, and what the real secret to wealth is
- Why feeling like a failure is a great place to start your success story
- The way to gain world-beating levels of focus, even if you normally struggle to concentrate

"This book provides a wealth of information on how to improve your thinking and your life. It is difficult to summarize the information provided. When I tried, I found I was just listing the information provided on the contents page. To obtain the value provided in the book, you must not only read and understand the provided information, you must apply it to your life."

NealWC - Reviewed in the United States on July 16, 2023

"This is an inspirational read, a bit too brainy for me as I enjoy more fluid & inspirational reads. However, the author lays out the power of thought in a systematic way!"

Esther Dan - Reviewed in the United States on July 13, 2023

"This book offers clear and concise methods on how to think. I like that it provides helpful methods and examples about the task of thinking. An insightful read for sharpening your mind."

Demetrius - Reviewed in the United States on July 16, 2023

"Exactly as the title says, actionable steps to guide your thinking! Clear and concise."

Deirdre Hagar Virgillo - Reviewed in the United States on July 18, 2023

"This is a book that you will reference for many years to come. Very helpful and a brain changer in you everyday life, both personally and professionally. Enjoy!"

Skelly - Reviewed in the United States on July 6, 2023

Bonus Nr. 2

Our Bestseller

The Art Of Game Theory

Total Value: $9.99

If Life is a game, what are the rules? And more importantly... Where are they written?

Here's just a fraction of what you'll discover inside:

- When does it pay to be a selfish player... and why you may need to go inside a prisoner's mind to find out
- How to recognize which game you're playing and turn the tables on your opponent... even if they appear to have the upper hand
- Why some games aren't worth playing and what you should do instead

"Thanks Wisdom University! This book offers simple strategies one can use to achieve things in your personal life. Anyone of average intelligence can read, understand and be in a position to enact the suggestions contained within."

David L. Jones - Reviewed in the United States on November 12, 2023

"Haven't finished it yet, but what I've gone through so far is just incredible! Another great job from this publisher!"

W. S. Jones - Reviewed in the United States on October 12, 2023

"A great book to help you through difficult and complex problems. It gets you to think differently about what you are dealing with. Highly recommend to both new and experienced problem solvers. You with think differently after reading this book."

Thom - Reviewed in the United States on October 18, 2023

"I like this book and how it simplifies complex ideas into something to use in everyday life. I am applying the concept and gaining a lot of clarity and insight."

Ola - Reviewed in the United States on October 18, 2023

"The book is an excellent introduction to game theory. The writing is clear, and the analysis is first-rate. Concrete, real-world examples of theory are presented, and both the ways in which game theory effectively models what actually happens in life is cogently evaluated. I also appreciate the attention paid to the ethical dimensions of applying game theory in many situations."

Amazon Customer - Reviewed in the United States on October 8, 2023

Bonus Nr. 3 & 4

Thinking Sheets

Break Your Thinking Patterns

&

Flex Your Wisdom Muscle

Total Value Each: $4.99

<u>A glimpse into what you'll discover inside:</u>

- How to expose the sneaky flaws in your thinking and what it takes to fix them (the included solutions are dead-simple)
- Dozens of foolproof strategies to make sound and regret-free decisions leading you to a life of certainty and fulfillment
- How to elevate your rationality to extraordinary levels (this will put you on a level with Bill Gates, Elon Musk and Warren Buffett)
- Hidden gems of wisdom to guide your thoughts and actions (gathered from the smartest minds of all time)

Here's everything you get:

✓ How To Train Your Thinking eBook	**($9.99 Value)**
✓ The Art Of Game Theory eBook	**($9.99 Value)**
✓ Break Your Thinking Patterns Sheet	**($4.99 Value)**
✓ Flex Your Wisdom Muscle Sheet	**($4.99 Value)**
✓ All our upcoming eBooks	**($199.80* Value)**

Total Value: $229.76

Go to wisdom-university.net for the offer!

(Or simply scan the code with your camera)

*If you download 20 of our books for free, this would equal a value of 199.80$

WHAT READER'S ARE SAYING ABOUT WISDOM UNIVERSITY

"Wisdom University embodies an innovative and progressive educational approach, expertly merging deep academic insights with contemporary learning techniques. Their books are not only insightful and captivating but also stand out for their emphasis on practical application, making them a valuable resource for both academic learning and real-world personal development."

—*Bryan Kornele, 55 years old, Software Engineer from the United States*

"Wisdom University's works provide a synthesis of different books giving a very good summary and resource of self-help topics. I have recommended them to someone who wanted to learn about a topic and in the least amount of time."

—Travvis Mahrer, BA in Philosphy, English Teacher in a foreign country

"I have most of the ebooks & audiobooks that Wisdom University has created. I prefer audiobooks as found on Audible. The people comprising Wisdom University do an excellent job of providing quality personal development materials. They offer value for everyone interested in self-improvement."

—Neal Cheney, double major in Computer-Science & Mathematics, retired 25yrs USN (Nuclear Submarines) and retired Computer Programmer

"WU is a provider of books regarding mental models, thought processes, organizational systems, and other forms of mental optimization. The paradigmatic customer likely is to be someone in an early- to mid-career stage, looking to move up the ranks. Ultimately, though, the books could be of use to everyone from high school students to accomplished executives looking for ways to optimize and save time."

—Matthew Staples, 45, Texas (USA), Juris Doctor, Attorney

"I have been reading books from Wisdom University for a while now and have been impressed with the CONDENSED AND VALUABLE INFORMATION they contain. Reading these books allows me to LEARN INFORMATION QUICKLY AND EASILY, so I can put the knowledge to practice right away to improve myself and my life. I recommend it for busy people who don't have a LOT of time to read, but want to learn: Wisdom University gives you the opportunity to easily and quickly learn a lot of useful, practical information, which helps you have a better, more productive, successful, and happier life. It takes the information and wisdom of many books and distills and organizes the most useful and helpful information down into a smaller book, so you spend more time applying helpful information, rather than reading volumes of repetition and un-needed filler text.

—Dawn Campo, Degree in Human psychology and Business, Office administrator from Utah

“I'm a subscriber of Wisdom University for over a year now. I would recommend Wisdom University books to anyone who wants to improve their understanding of cognitive behavioural therapeutic principles.”

—Sunil Punjabi, Maharashtra (India), 52, PhD, Psychologist

“I wanted to read some books about thinking and learning which have some depth. I can say "Wisdom University" is one of the most valuable and genuine brands I have ever seen. Their books are top-notch at kindle. I have read their books on learning, thinking, etc. & they are excellent. I would especially recommend their latest book "Think Like Da Vinci" to those who want to have brilliant & clear thinking.”

—Sahil Zen, 20 years old from India, BSc student of Physics

“I came to know about Wisdom University from the Amazon Kindle. There were recommendations for some of the Wisdom University books. Found every book very interesting. I really loved it. Subscribed for the free material which was delivered right into my inbox. Since then, I have been a fan. I couldn't buy the books... since am in a situation. But as soon as I get a sufficient amount, I plan to purchase some nice titles that piqued my interest. I recommend the books to everybody who wants to live a life free from all sorts of mental blocks that reflect in real life. These books are definitely the lighthouse, especially for those crawling through the darkness of ignorance. I wish Wisdom University all the best.”

—Girish Deshpande, India, 44, Master of Veterinary Science, working as an Agriculturist

INTRODUCTION

Imagine you are poised at the edge of a vast forest. Deep within lies the blueprint for creating not only thriving communities but also a sustainable world. The challenge? Choosing the right path to navigate the forest depths.

Sustainability champions and community builders around the world are driven by a shared vision. They see a world where every individual is not just a bystander but an active participant in shaping their environment, making informed decisions, and contributing to the collective good.

The aspiration is clear: thriving communities within ecosystems that are balanced and sustainable.

However, the path to the vision is riddled with complexities. The nuances of local cultures blend and clash with broader global sustainability issues. Mobilizing resources requires collective responsibility. The task is enormous and intricate. Traditional approaches have missed the deeper, interrelated dynamics underpinning our world.

Enter Systems Thinking.

The Tale Of Maya's Orchard

A quaint town named Selenia lay in the heart of the valley and was known for its lush apple orchards. Mava, a young agronomist, returned to her hometown after studying sustainable agriculture in a nearby city. Coming home, she noticed a concerning trend: the apple yield decreased year over year, and the fruits were smaller, with less flavor.

The town's residents had met to discuss the problem. They were implementing an immediate solution of using more fertilizer and pesticides, thinking the apple yield would increase and the apple quality improve.

Mava, however, was determined to find a lasting solution and decided to investigate. Rather than focusing narrowly on the apple trees, she looked at the entire ecosystem surrounding the orchard.

She learned that the local bee population, critical to pollinating the apple blossoms, had declined. The increased use of pesticides was inadvertently harming the bees.

Furthermore, she observed that the soil was deteriorating. The continuous use of fertilizers had stripped the soil of its natural nutrients and beneficial microbes, making the trees more susceptible to disease.

Armed with her knowledge, Mava proposed a systems thinking approach to the town's council. She introduced them to organic farming practices.

She advocated for planting wildflowers around the orchards to attract and support bees and other helpful insects. She explained how they could use composting and crop rotation to rejuvenate the soil naturally.

Over the next few years, Selenia's orchards began to thrive, and the apple yields increased. The fruits also regained their luscious color and flavor. Bees and butterflies darted through the orchards, reflecting the balance that had returned.

Mava's holistic approach to Selenia's challenge embodies the principles of systems thinking. It underscores the importance of understanding the interconnectedness of nature and the ripple effects of actions.

What Will This Book Teach You?

This book is more than just a guide. It is a transformative journey into the world of systems thinking, tailored for those eager to make an honest difference. Here is what the book promises:

A new lens: Embrace a perspective that sees communities and ecosystems not as separate entities, but rather as dynamic, interconnected systems.

Clear vision: With a metamorphic understanding, begin to strategize holistically, inclusively, and sustainably.

Deep comprehension: Explore beneath the surface to understand the root causes of challenges, moving beyond visible events and symptomatic solutions.

Empowerment: Lead others to engage in systems thinking, fostering collective responsibility and proactive global guardianship.

Resource mobilization: Learn to engage your inner circle in community causes that make a difference. Expand that small circle to influence larger issues in broader regions.

While you may think you don't have time to learn an overwhelming amount of new knowledge, you will be surprised at how easily you can learn using simple exercises and activities that take very little time. While systems thinking is complex, we break it down into small steps, real-world examples, and fun learning activities.

The cost of not learning systems thinking is more than the cost of learning it. Coming away from the learning with a significant ability to improve the world will pay huge rewards in the future. You might wonder, can I even do this? We will learn about Wangari Maathai and her small group of Kenyan women who used systems thinking to save the Kenyan forest. Yes, you can make a difference.

You may feel that your life is fine as it is, but when you read the many stories of success using systems thinking in large and small challenges, I think you will want to be a part of the movement.

For those passionate about shaping a brighter, more sustainable future, integrating systems thinking is more than a new approach; it is a paradigm shift–a shift that can potentially redefine our ecosystems, communities, and individual lives.

Who Is The Author?

My name is Susan Ferebee. I have my Ph.D. in Information Systems and I have worked many years in the industry, as well as teaching for universities in Information Technology, Cybersecurity, Critical Thinking, Systems Analysis and Design, Project Management, and Software Development.

In my years working for large technology organizations, I was a senior project manager for several projects where systems thinking was applied. I wrote process simulation programs to see systems in action and to depict their connections to other systems. I am passionate about opening everyone's eyes to the lens of systems thinking, as I know how enlightening it is for your future career, community work, and even in understanding and nurturing your family.

Your Path To Systems Thinking Mastery And A Sustainable Future Begins Here

This book will tell you the origins of systems thinking and how it theoretically emerged as a way of thought. However, the book aims to teach you how to put systems thinking into action in small scenarios and in large social issues. You might feel your world is too small to apply systems thinking to—it is not. Your journey through this book will result in your being a lifelong systems thinker, applying this method to everything you observe around you!

As you embark on this exciting adventure, prepare to see the world through a new lens, uncover hidden pathways to understanding, and unlock the immense potential for collaborative, systemic action.

Welcome to the life-changing power of systems thinking. Are you ready to move to a balanced, sustainable future in all aspects of your life? *Keep reading!*

Susan Ferebee

1

WHAT EXACTLY IS A SYSTEM?

AND WHY SHOULD YOU CARE?

Trees During Daytime by Gustavo Queiroz

A mesmerizing scene unfolded before Isabelle in the heart of the dense forest. Rays of sunlight pierced the thick canopy, weaving intricate light patterns and shadows across the forest floor.

As she continued deeper into the forest, a captivating sight snapped into focus—a vibrant ecosystem thriving around the base of a towering pine. Curious, Isabelle knelt to examine the intricate dance of life at play.

It was a system operating long before her arrival, and she was about to unravel its secrets.

A fallen tree trunk lay partially covered in moss at the center of this microcosm. Isabelle explored the symbiotic relationship between the moss and the decaying wood.

The moss clung to the wood's surface, drawing moisture from the air and creating a lush, emerald carpet. As the moss grew, it released organic matter, nurturing the soil.

Around the mossy trunk, a network of fungi stretched its delicate threads. These fungi formed a hidden partnership with the towering pine tree, extending their filaments into its roots. In exchange for sugars, the fungi helped the tree absorb minerals and water from the soil.

As Isabelle's gaze shifted outward, she saw bees darting among the leaves, pollinating flowers, and chipmunks scurrying, cheeks swollen with small berries. In a cascade of interactions—pollinators aided reproduction, herbivores spread seeds, and predators kept the balance.

Amidst the rich foliage, ivy tendrils reached out, using the tree as support to climb skyward, pursuing the sun. This showcased nature's ingenuity through an adaptation that allowed it to thrive by exploiting available resources.

Enchanted by this interconnected web of life, Isabelle realized that nature was a symphony of systems. Each system element seamlessly wove into the next, creating a harmonious dance of growth, balance, and life.

Isabelle left the forest with a newfound appreciation of the systems operating in the natural world. She comprehended that just as the forest held its systems to be unveiled, the designs of our everyday lives were equally captivating, awaiting discovery and understanding.

Intertwined Mysteries: Beyond The Obvious

Through Isabelle's story, we glimpse the intricate workings of systems through the lens of nature. Isabelle's experience illustrated the fundamental essence of systems—a dynamic interplay of interconnected elements. Each element's behavior influences the whole.

But what exactly is a system? Why is understanding it so important in navigating our world's complexities?

Systemic structures exist everywhere, from the natural world to human creations. They can be microscopic to global.

They can be as simple as a swinging pendulum influenced by gravity, intricate like the forest, or highly complex, like the world's climate.

A system is an emergent pattern that develops from individual interactions. It is more than the sum of its parts and represents collective behavior governed by simple rules[1].

Consider these systems:

- a flock of birds in synchronized flight
- a city
- the Earth
- a family
- your body
- an automobile
- the electric grid
- a school
- your city government

For each of these systems, consider the elements, the connections, the interactions, and the effects of those connections and interactions. Consider how the system changes over time. The arrangement of a system's elements defines its structure. Its design determines its function, and how it behaves represents how it changes over time.

A good example is thinking about an Arctic glacier melting and the impact it has on worldwide rising sea levels. What is the further effect that now occurs in coastal cities and wildlife?

Begin to observe the systems around you. What are the elements that comprise the organized structure? How do those elements interact? How do these systems contribute to the challenges in our world today? Everyday systems you might interact with include:

- the self-checkout at the grocery
- the drive-through system at your favorite fast-food restaurant
- the traffic control system in your city

- the local library
- the bank's drive-up ATM

In the following sections, you will see how to dissect system elements and unravel the interactions in systems. Open your mind to the complexities of systems!

So, what makes something a system? There are four key things: components, connections, function, and purpose, and change over time[2]. A system is a group of things that work together to achieve a common goal.

For example, your body is a system of organs, tissues, bones, and cells. These components connect through biological processes, like the lungs exchanging oxygen and carbon dioxide, the heart pumping blood, and the nervous system transmitting signals. The purpose of your body is to maintain life and provide optimal functioning. And as you grow and change, your body changes too.

A subsystem is a minor system that is part of an extensive system. For example, the circulatory system is a subsystem of your body. It comprises the heart, blood vessels, and blood. The circulatory system aims to deliver blood, containing nutrients and oxygen throughout your body and eliminate bodily waste.

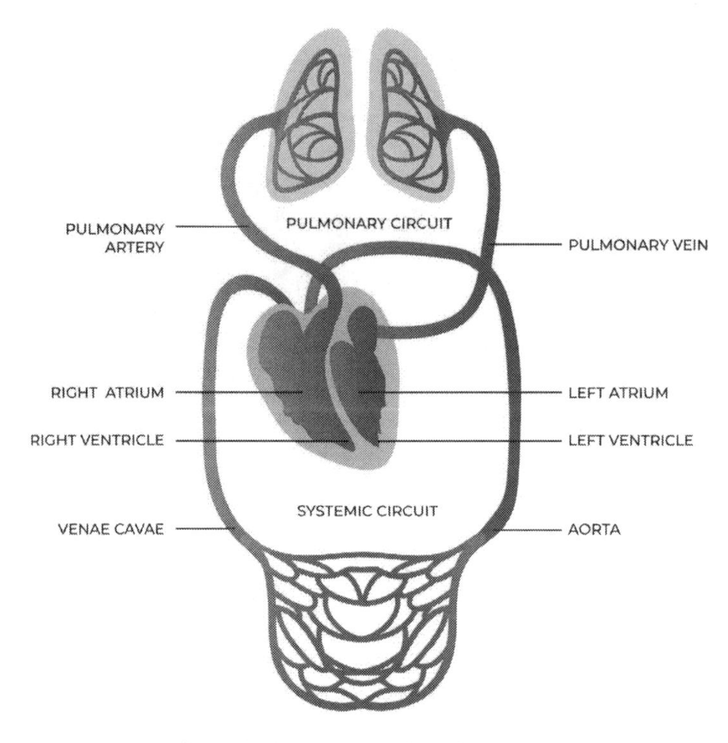

CIRCULATORY SYSTEM

Circulatory System Infographic by Freepik[3]

In the circulatory system, you can see the interaction of the heart and lungs as blood is transported through the body carrying nutrients, waste byproducts, and oxygen. The elements are the heart, blood, and blood vessels.

There are four concepts to help you understand what a system is doing: a) boundary, b) openness, c) self-containment, and d) adaptation[4].

A boundary or edge marks where the system ends and its surrounding environment begins. A single cell is an organized structure; the cell wall is its boundary. The body of a car represents the edge of the automobile system. A person's skin is the boundary for their entire body.

Different boundaries exist. The skin of the human body is physical. There can be functional perimeters. Think of a fast-food restaurant with perimeters between the inside counter and the drive-through ordering system.

The second principle that Collins[5] discusses is the concept of open and closed systems. An open system exchanges energy and material with its surrounding environment and adapts to environmental changes. A closed system does not interact with its surroundings and is self-contained, with all elements and interactions needed.

All living organisms are open as they depend on exchanging food, water, carbon dioxide, and oxygen with the external environment. Humans move through their external environment, continually interacting with it.

Seedling in Garden Soil by Karolina Grabowska

An excellent example of a closed system is a watch with multiple interacting parts in a sealed environment to protect it from dust and water.

There are different types of systems:

Biological - a living organism, such as a human, animal, or plant.

Economic - comprising currencies, products, and services.

Computer - a collection of hardware and software working together to process information.

Social - a group of related people interacting, such as a family, neighborhood, gang, club, or community.

Political - a way of organizing the government and exercising power to make decisions for the public.

Natural - for example, ecosystems or weather systems.

Elemental Enigmas

The elements or components of a system are the individual parts that make up a whole system. These separate parts interact with one another to complete the purpose of the system.

The elements of a system can be tangible or intangible. Consider a family. Actual components are the family members. Intangible elements include loyalty and belonging. The mother of the family is an element of the larger family system.

If you think about a government system, tangible components are the people in the political offices. Abstract elements are government policies. As political figures change, policies often vary as well. Changes to one element usually result in changes to other elements.

System components or elements are called stocks, the measurable objects in a system. If we are looking at a hospital system, stocks would include the number of patients, nurses, doctors, hospital departments, types of disease, stress levels of doctors, nurses, and patients, and government regulations. A stock can be living, non-living, or conceptual[6].

Stocks show the state of a system and form the foundation for actions, while adding memory and passivity to the system. They provide continuity between prior and current conditions, cause delays and disrupt flows, and tend to be represented as nouns (naming things or status).

Flows can influence stocks. They are the inputs to and outputs of the system, and are represented as verbs (describing activities and actions)[7].

You can also think about the more extensive system any other system might be a part of. The family is an element within the larger population system. A city government is a part of the country's government. Every system has components, but systems can become elements within another more extensive system[8].

Henshaw et al[9]. refer to these interconnected systems as a System of Systems (SoS) and state that the SoS has a distinct ability that, alone, none of the individual systems could accomplish. The U.S. Department of Defense (D0D) describes engineering an SoS as synthesizing the abilities of a group of systems into an SoS with capacities greater than the sum of the individual system's capabilities[10].

An SoS is complex and often distributed, meaning the varying constituent systems might be in different places connected through technology. A good example is the global air traffic control system, which combines communication, radar, and navigation systems.

Think of the Earth system. Within the Earth, are several continents, several oceans, and the North and South Poles. At the more abstract level are the weather, the ocean tides and currents, and airstreams.

Earth Satellite by 12019

Each continent has subsystems of countries, forests, rivers, and plains. A country has further subsystems of states and cities. You can see the almost never-ending levels of systems and subsystems that can be identified.

Your car, for example, is part of the transportation system, and your computer becomes part of the internet when you connect. The more extensive system provides the context for the system elements. It defines the boundaries and the relationships between the system elements[11].

Ties That Bind

To achieve the purpose of the system, the system elements must be connected and interact[12]. The system elements are not independent of each other. They are interconnected and interdependent. The changes in one system element can affect the other system elements. For example, if the engine in a car breaks down, the vehicle will not be able to move. This is because the engine is a system element interconnected with other system elements, such as the wheels and the transmission.

In a vast system, like the United States, when a flood, hurricane, or tornado destroys a region, transportation, energy, and food supplies are disrupted for that area and many other regions. Some interconnections are less visible or even invisible.

For example, when a storm destroys cities and farms, it can significantly affect the microelements in the system. Microelements are nutrients required by plants and animals to grow and thrive. They are often found in small amounts in the soil, but storms can wash them away or damage the soil structure, making it difficult for plants to absorb them. This can result in a lack of fundamental nutrients for both plants and animals, which can have a descending impact on the entire ecosystem.

An invisible connection is information flow through a computer system. Hidden interconnections can exist between a child and their parents when the child favors one parent or between bosses and employees.

Think of your workplace as a system. Consider what the elements of the system are and how they are interrelated. Your company includes employees, buildings, technologies, policies, rules, and an intangible organizational culture. Think about whether the elements of your organization's system are promoting or hindering business success.

We see the importance of elements and their connections and interactions in looking at the societal issue of drug addiction. Over the years, it has become obvious that addicts and their families rarely succeed in curing addiction. This is because addiction must be seen within the larger societal system it is part of[13].

Silent Beacon: The System's True North

A group of elements, even when interconnected in some way, does not comprise a system unless that system has a purpose and function[14]. The purpose of a system is why it exists. The function of a system is how it achieves its purpose. For example, the purpose of a car is to transport people and goods. The function of a car is to convert fuel into energy that moves the car's wheels.

Here is a fictional story that illustrates the importance of a system's purpose.

Aqualink: The Tale Of Bridgewood's Waters

In a bustling city named Oxbridge, a grand library was known as the "Library of Systems." This library was unique because it didn't just house books; it held intricate models of various systems designed by the brightest minds of Oxbridge. Each system had a purpose, and a story reflected each intent.

One day, a young man named Jordan visited the library. He was particularly interested in understanding the purpose of a newly introduced AquaLink system. He approached the librarian, Mr. Elevate, and asked, "What is the story behind AquaLink?"

The librarian smiled and began with:

"In a distant part of Oxbridge was a village named Bridgewood. The villagers depended on a single well for their daily water needs. However, as the town grew, the single well could not meet the demand. Villagers had to wait in long lines, and often, the well would run dry by the end of the day."

"Seeing the villagers' plight, Elio, an engineer, decided to design a system to solve the problem. He observed the village's layout, daily water consumption, and nearby water sources. After months of research, he came up with AquaLink."

"AquaLink was a network of interconnected water tanks placed strategically throughout Bridgewood. These tanks were connected to various water sources, including the well, a nearby river, and rainwater harvesting systems. The system used sensors to monitor the water level in each tank and would automatically redirect water from abundant sources to tanks running low."

"The beauty of AquaLink was its adaptability. During the rainy season, it prioritized rainwater. In the dry season, it balanced between the river and the well. It also had filters to ensure the villagers received clean water regardless of the source."

"With AquaLink in place, the villagers no longer had to wait in long lines. Water was available round the clock and year-round, and the system ensured equitable distribution to all village areas."

Jordan's eyes sparked with understanding. "So, AquaLink's purpose was to ensure a consistent and clean water supply to all villagers, adapting to changing conditions."

"Exactly," replied Mr. Elevate. "Every system in this library has a purpose, and behind each purpose, there's a story that reflects its essence."

Jordan left the library that day with an appreciation for systems and the stories that define them. He realized that understanding the story behind a system is crucial to genuinely grasping its purpose.

<u>Systems' Mysterious Motives</u>

An organization has a clearly stated purpose in its vision, mission, and goals. The climate system, however, may have a less obvious purpose. Watching a system over time reveals its behavior and purpose[15]. Consider the following systems that have ambiguous purposes:

- The United States Congress

- The Stock Market
- Facebook

The purpose of Congress is to enact laws that influence the lives of U.S. citizens. Congress serves as the voice of the people within the government[16]. However, Congress is a bureaucracy that serves diverse and contradictory objectives, bound by rigid rules and procedures that often obscure the purpose of the congressional system.

The stated purpose of the stock market is a system where company ownership shares can be bought and sold. That is the stock market's function. Its purpose, however, needs to be clarified. One purpose is for organizations to raise needed capital. Another purpose is for investors to profit. The multiple purposes can be contradictory.

A social media system like Facebook aims to connect everyone globally. The purpose was to empower individuals to build a community that drew the world closer[17]. However, over time, Facebook's leadership behavior morphed the purpose to include advertising revenue as a secondary and conflicting purpose.

Whispers Of The Winds Of Change

The purpose and function of a system can evolve naturally over time or, sometimes, become something never intended. Different elements of systems may develop divergent purposes[18].

According to Rutherford[19], changing the interrelationships and connections between system elements has a greater impact on the whole system than changing the system elements. Changing the purpose of a system has the greatest effect.

The telephone system has changed its purpose and function significantly since the first phone was developed by Alexander Graham Bell. Its original purpose was to allow people to communicate across long distances. In achieving this purpose, there were changes from switchboard operators to automatic telephone exchanges to mobile connections—not requiring people to be in a specific location to make a call.

The internet changed the network over which calls could be made and contributed to changing the purpose of the phone (now smartphone)—from communicating with people at long distances to combining this purpose with serving as a computer and camera. The purpose of a smartphone is to give users a mobile device for making and receiving voice calls, sending and receiving text messages, searching the internet, listening to music, taking photographs, using apps, playing games, and connecting with friends and family on social media.

Conclusion

We learned from Isabelle's journey into the forest that the essence of a system, like the forest, is a dynamic interplay between interconnected elements. Each element's behavior influences the whole.

To continue understanding systems, learn to recognize them all around you. In the next chapter, we will further define and demonstrate systems, subsystems, and the communication and interactions between them.

As you become able to recognize systems as part of a larger whole, you will be ready to progress to the following chapters on systems thinking and you will learn how to apply systems thinking to solve small and large problems

Action Steps

Tomorrow, as you navigate your daily life, identify at least two systems that you interact with.

Describe the elements (stocks) of the system.

Describe the interconnections and interrelationships (flows) of the system.

Determine the primary purpose of the system.

Does the system have conflicting purposes?

Chapter Summary

1) Systems are made up of elements.

2) The elements in a system are interconnected and interrelated.

3) To be considered a system, the interrelated and interconnected elements must have a purpose.

4) A system's purpose might be singular, but can also be ambiguous and there might be conflicting purposes.

5) Systems change over time.

6) A System of Systems (SoS) comprises several systems synthesized into a single larger system with a distinct ability that alone, none of the individual systems could accomplish.

2

(EXTRA)ORDINARY ENCOUNTERS WITH SYSTEMS

HOW SYSTEMS WORK IN THE WORLD AROUND US AND HOW TO WORK WITH THEM

Edward Lorenz, in 1972, coined the butterfly effect. The term refers to the idea that a butterfly flapping its wings in South America could start a tornado in Kansas. As outlandish as this may seem, the underlying principle is profound: Small changes can lead to disproportionately large effects in a system.

The story begins in the sixties with Lorenz, a Massachusetts Institute of Technology meteorologist. He was working on a set of mathematical equations to predict weather patterns. To save time, he entered a shortened number into his computer program one day, expecting a similar outcome. To his amazement, the results were drastically different—a minor change had produced a completely different weather pattern[1].

The Butterfly Effect In Action

Butterfly Effect Waves by Gerd Altmann

In the mid-2000s, a mysterious phenomenon made headlines worldwide: Colony Collapse Disorder (CCD)[2]. Beekeepers reported that their honeybee colonies were dying off at an alarming rate. Hives that were once buzzing with activity were suddenly empty, with no clear explanation for the mass disappearances.

The implications of this were far-reaching. Honeybees are crucial in pollinating many crops that comprise a significant portion of the world's food supply. From citrus in California to apple orchards in China, the decline in bee populations threatened global food production systems.

As scientists scrambled to understand the cause of CCD, a web of interconnected systems emerged. Pesticides, particularly a class known as neonicotinoids, were identified as a potential culprit. These chemicals, designed to protect crops from pests, inadvertently affected the nervous systems of bees, disorienting them and making it difficult for them to return to their hives.

But the pesticides were just one piece of the puzzle. Modern agricultural practices favored vast monocultures reducing the diversity of plants available for bees to feed on. This lack of variety weakened their immune systems, making them more susceptible to diseases and pests[3].

The impact on the agricultural system was immediate and devastating. In regions heavily affected by CCD, crop yields plummeted. The price of certain foods, like almonds, skyrocketed. Farmers, facing reduced incomes, struggled to keep their operations afloat[4].

The ripple effects have not stopped. The economic system felt the strain as food prices fluctuated. In some areas, job losses in the agricultural sector led to increased unemployment rates. The ecological system, too, was affected as the reduced number of bees impacted the reproduction of wild plants, leading to decreased biodiversity[5].

The story of CCD serves as a stark reminder of the fragile interdependence of the world's systems. A disruption in one can set off a chain reaction with far-reaching and unpredictable consequences.

It underscores the importance of understanding these connections and the need for sustainable practices that consider the broader impact on the environment and society.

Systems Through Time's Lens

System definitions have evolved for many centuries. A Russian physician, Alexander Bogdanov (1873-1928), derived a system theory called "tektology," which translated to a science of structures. Tektology revealed how nature and human activity were organized, examining the whole of connections between system components[6].

Ludwig von Bertalanffy (1901-1972) was a biologist in Vienna who developed the general systems theory. According to Bertalanffy, in the general systems theory, living things are open systems because they require a continuous input of matter and energy from the external environment to survive[7].

Norbert Wiener (1894-1964) formulated an approach to system communication and control called cybernetics and defined it as animal and machine control, feedback, and communication[8]. Wiener described cybernetics as an attempt to bring together the study of electromechanical and biological systems which share the principles of feedback, control, and communication[9]. Wiener speculated that intelligent action resulted from feedback received by the human body and believed that this intelligent behavior could be replicated by machines[10].

James Grier Miller (1916-2002) formulated the Living Systems Theory and defined living systems as self-organized, open, and interacting with the environment through the exchange of information, energy, and matter. Unique to Miller's theory was an eight-level system hierarchy[11]:

1. Cells
2. Organs
3. Organisms
4. Groups
5. Organizations
6. Communities
7. Societies
8. Supranational

A more modern theory, Social Systems Theory, was derived by Talcott Parsons (1902-1979), Walter Buckley (1922-2006), and Niklas Luhmann (1927-1998). In the Social Systems Theory, it is believed that every social contact is a system making up a larger society system.

Piryankova[12] expanded the definition of a system, describing it as a complex arrangement of interconnected, evolving subsystems. This aligns with the concept of a System of Systems (SoS) that we examined in Chapter 1.

The Hidden Harmony Of Systems

Stack of Stones Outdoors

Several characteristics contribute to the continuous, successful functioning of a system: 1) self-organization, 2) hierarchy, and 3) resilience[13]. These characteristics are found in machine, biological, and social systems.

Systems can exist in a static form until threatened, and once threatened, self-organizing behavior allows the system to adapt to an external threat. For example, if a non-native species is introduced into an ecosystem, it might cause diseases in existing species. A self-organizing response might be the native species recognizing the invasive species as a food source, controlling its population.

In a computer system, self-organization is not inherent within the system but adaptation occurs in tools that are installed to protect the system. For example, anti-malware software adapts with continuous updates to databases to include new malware signatures. Other tools rely on machine learning to understand previous attacks and adapt based on that learning.

As systems adapt, hierarchal structures form within the system elements to ensure efficient use of resources. As the system adapts, it strengthens and becomes resilient.

Self-Organization

A self-organizing system can change its internal organization and function in response to circumstances in its external environment[14]. The purpose of this self-organization is to survive external events. The self-organization can lead to a hierarchy of structure and behavior.

Do you remember the forest system from Chapter 1. What if that forest was threatened by an external influence that could destroy some of the forest elements? How could the forest communicate and self-organize to protect itself? The scientific research that demonstrates the amazing way that a tree's roots can distinguish themselves from the roots of other trees and recognize other species' roots and how trees communicate[15] is discussed below.

Trees With Roots Above Ground by Aneesh Aby

Suzanne Simard, in 1997, published results of her dissertation studies on forest tree communication[16]. In a 2016 interview with Diane Toomey of the Yale Environment 360, Simard discussed this continued research over many years, saying that trees not only transfer resources between themselves but also signal distress, defense, and kin recognition. The trees also develop an interdependent relationship with underground fungi[17].

The mycorrhizal (the combined Greek words *mykos* for fungus and *riza* for root) network described above is the underground network created when the filaments of mycorrhizal fungi connect with tree roots, thereby connecting the trees[18].

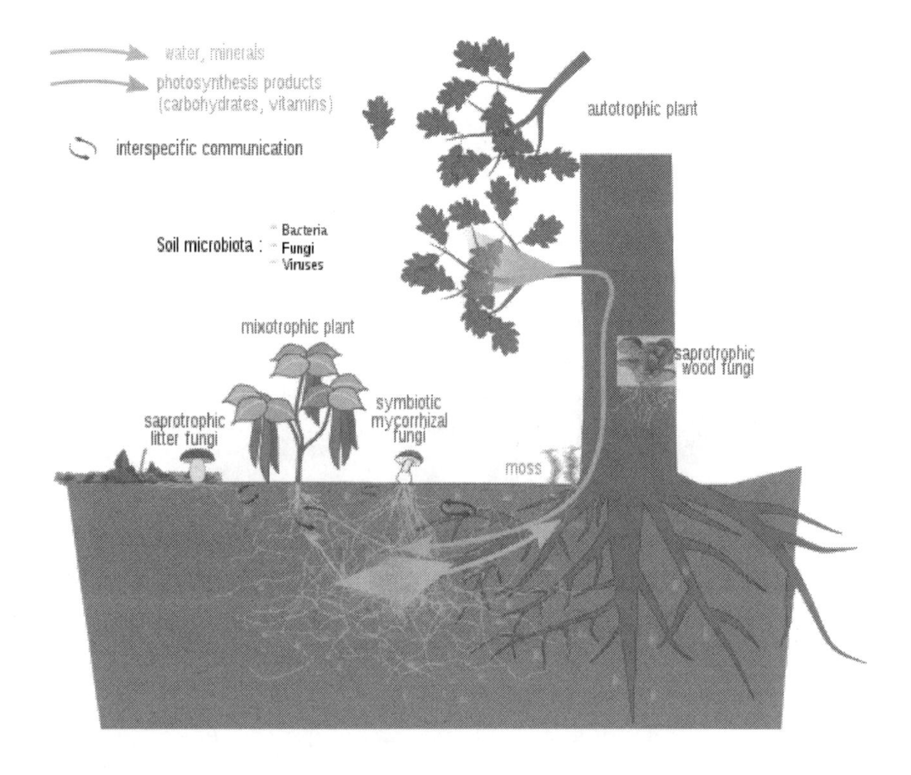

Mycorrhizal Network by Charlotte Roy, Salsero35, and Nefronus[19]

Merlin Sheldrake, an English plant scientist and expert in mycorrhizal networks stated, "You could imagine the fungi themselves as forming a massive underground tree, or as a cobweb of fine filaments, acting as a sort of prosthesis to the trees, a further root system, extending outwards into the soil, acquiring nutrients and floating them back to the plants, as the plants fix carbon in their leaves and send sugar to their roots, and out into the fungi. And this is all happening right under our feet[20]."

Simard explains that ecological conditions determine how many nutrients are exchanged and in which direction. As conditions change, the forest system self-organizes to adapt. She describes a forest with Douglas fir and Birch trees. When the Douglas fir becomes heavily shaded during summer months, a greater amount of excess carbon stored by the Birch is transferred to the Douglas fir. In reverse, in the fall, when the Birch trees lose their leaves, and the Douglas fir now has excess carbon, the greater exchange of carbon flows back to the Birch trees.

The largest trees have the largest root systems, and therefore, have the most interconnected roots. Simard also discussed "mother trees" and their ability to recognize their own offspring, seedlings that regenerate near them. This was proven with DNA testing, according to Simard. These mother trees do recognize their kin and nourish their own kin.

When there is a threat to the forest like the mountain pine beetle which kills pine and spruce trees, how does the mycorrhizal network respond? Does self-organizing occur? According to Simard, research showed that as the mountain pine beetle attacks continued, the mycorrhizal network diversity decreased and the defense enzyme in new seedling trees changed to focus on the threat. Even with this change in fungal diversity, the mycorrhizal network adapted to nourish and protect the seedlings and regenerate a new stand of trees[21].

Hierarchy

Often, hierarchical structures and relationships are formed when systems self-organize to survive an external influence[22].

A hierarchical structure exists in the human body where a cell in the lung is a subsystem of the lung, which is a subsystem of the circulatory system, and finally, which is a subsystem of the human body.

In our forest example, you can think of the fungi as a subsystem of the roots which were a subsystem of the trees, with each individual tree being a subsystem of the forest. The hierarchy also consists of different tree varieties and young and old trees.

Hierarchical structures ensure that resources are efficiently used, with each layer performing a specific function. Minimizing resource waste maximizes resilience and adaptability. In a hierarchical structure, interrelationships within subsystems are stronger than interrelationships between subsystems.

Ant colonies provide a perfect example of hierarchy in a system. Each ant has a specific role determined by a clear hierarchy: the queen, soldiers, workers, and drones. The queen focuses solely on laying eggs to ensure the colony's future.

Worker ants gather food and tend to the young, while soldier ants protect the colony from threats. This hierarchical structure ensures efficiency, a clear division of labor, and the colony's survival against external challenges.

Resiliency

A system is resilient if it can be threatened and, after reacting to protect itself, springs back into its original form[23]. Within an evolving environment, the system can persist. Resilience depends on numerous feedback loops that react to environmental changes to restore normalcy. At the core of resilience is a simple concept: things change, and systems must respond[24].

Resiliency occurs through the presence of these key system attributes:

Redundancy - backup systems or multiple paths to a system element ensuring that another can take over if one system part fails.

Modularity - partially independent elements compose the system so if one element is threatened, the entire system is not under the same threat.

Feedback loops - system sensors detect changes in the external environment and adjust accordingly, providing the system with adaptability.

Robustness - inherent strength of a system (size, material composition, structure).

Self-organization - the system can reorganize from within its boundaries to respond to environmental changes.

Learning and memory - the system learns from inputs from feedback loops and remembers successful responses. Memories inform future responses.

Rapid response and recovery - how fast a system can respond, adapt to, and recover from threats determines its resiliency.

Buffer capacity - a system's capability of absorbing disturbance without fundamentally changing.

Diversity - a variety of elements and processes that respond to different challenges so there is no single point of failure.

The Resilient Tree

Root Tree by FelixMittermeier

We have discussed the forest system, showing the interconnected elements and its ability to self-organize to survive. Another example of why forest systems are strong is the resilience of a key forest element, the tree. Trees have long been a symbol of resilience and demonstrate hierarchy, feedback loops, subsystems, interconnection, communication between subsystems, and learning from the past.

The roots and branches of a tree are sensors and subsystems that gather information and resources from the environment.

The tree trunk is the command center and main structural element. As nutrients and water are gathered from the roots, the trunk directs those nutrients to the branches.

The tree is hierarchical. The branches, for example, subdivide into smaller branches and twigs, each with its own feedback loop, tied ultimately to the trunk. A similar hierarchy exists in the roots.

Leaves and roots serve as sensors and provide feedback, allowing the tree to adjust to changing environmental conditions.

Growth rings record the historical development of the tree, providing memory from which the tree learns and adapts.

Suboptimization

While self-organization, hierarchy, and resiliency maintain the system's efficient performance, systems do malfunction. One cause of a system malfunction is suboptimization, where one subsystem dominates, disrupting the hierarchy[25].

In the human body, for example, cortisol is released in a high-stress situation preparing the body to respond to an immediate threat. However, some people live in continuous stress situations causing too much cortisol to be released over time. This continuous cortisol release diminishes the immune system, causes weight gain, and raises blood pressure. An overactive subsystem is negatively affecting the larger system.

In an organization, decisions that optimize for the short term may diminish long-term results. Focusing on profit and ignoring sustainability could result in the company's ultimate demise.

Social systems also experience suboptimization. An example could be the transportation system which, in the United States, is optimized for individual convenience with most people driving their cars. This leads to heavy traffic and air pollution.

Beneath The Surface Of Societal Strains

As you begin to think in terms of systems, begin to also look at social systems and the flaws in those social systems that might reveal solutions to social problems. Characteristics of systems that might lead to issues are suboptimization, ambiguous boundaries, system delays, and ineffective feedback loops[26].

A Tale Of Two Patients

We can examine a hypothetical scenario for two individuals, Sara and Jane. They both fell seriously ill around the same time. Sara, a well-off executive, had a comprehensive health insurance plan, while Jane, a part-time waitress, had minimal coverage due to her employer's limited benefits and her inability to afford a better plan.

When Sara felt the first signs of her illness, she immediately consulted her primary care physician, who promptly referred her to a specialist. The specialist recommended a series of tests, which were swiftly approved by Sara's insurance. Within a week, Sara was diagnosed and began a treatment regimen tailored to her needs. Her insurance covered the majority of her medical expenses, and she could focus solely on her recovery.

Jane's journey was starkly different. Lacking a primary care physician, she first sought help at an urgent care clinic. The doctor there did his best but recommended seeing a specialist for her symptoms. However, Jane's insurance required a lengthy approval process for specialist visits. By the time she got the approval, her condition had worsened. The specialist recommended tests, but again, insurance delays meant valuable time was lost.

When Jane was finally diagnosed, she was presented with a treatment plan that was partially covered by her insurance. The out-of-pocket costs were astronomical. Jane had to make the heart-wrenching decision to opt for a less effective treatment because it was all she could afford.

The system flaw revealed in this story is the inequity in healthcare access and treatment based on one's financial status and insurance quality. In a system where profit often takes precedence, those without comprehensive coverage can face delays, receive subpar treatment, or be burdened with insurmountable debt, highlighting the urgent need for healthcare reform that prioritizes patient well-being over profits.

This issue reflects suboptimization because the health insurance subsystem dominates other elements of the social system. Decisions that had to be made by Jane and Sara were driven by their health insurance requirements. Implementing universal healthcare might balance the healthcare system and provide equally distributed quality healthcare.

You can also ask if a lack of diversity (a necessary attribute for system resilience) contributed to this healthcare inequity. Rather than having multiple options and paths toward quality healthcare, the options were limited or non-existent.

Were there ambiguous boundaries in the healthcare system? System delays existed in Jane's situation. Were there issues with feedback loops that might have improved Jane's ability to quickly access quality care? These are the types of questions to ask in systems thinking.

Systems Thinking - Healthcare Inequality

In the next chapter, we will explain and explore systems thinking in depth. But, Weinberg[27] offered three simple questions that you can start with now. Experiment with these simple questions to evaluate Jane and Sara's healthcare inequity.

Why do I see what I see?

Jane and Sara have different incomes and different health insurance options, so the end result of what you see is two diagnoses and treatment paths that led to Jane receiving subpar treatment compared to Sara.

Why do things stay the same?

In the United States, you might conclude that Congress's inability to compromise and enact healthcare reform is why the system does not change.

Why do things change?

Many factors lead to system change. In this particular situation, we could suggest that a cultural and social shift would need to occur so that people's health is valued over institutional profit.

Another systems thinking approach is to ask where are the leverage points in a system where a small change to one system element or process might significantly affect the overall system. Consider whether the leverage point here would be healthcare providers or health insurance providers. In which element would a minor change be likely to have a large impact?

The clear models used in systems thinking offer a new technique of understanding, revising, and testing how we understand the nature of the healthcare system. New approaches to intervention are offered that can improve everyone's health[28].

Conclusion

Lorenz's butterfly effect, where a minor change in one system element can result in a profound change in other parts of a system, was illustrated well in the Colony Collapse Disorder (CCD) that occurred in the 2000s.

The CCD demonstrated the importance of looking at a specific system within the larger system. Looking at the beehives and bees alone could not reveal how the monoculture agriculture practice that reduced plant diversity affected the bees' immune system, making them more highly susceptible to pesticides.

Learning to see systems within a larger whole is the key to developing a systems thinking approach. In the next chapter, we will define systems thinking and apply it to simple, intricate, and complex systems.

Action Steps

Consider a complex social issue you are familiar with (income inequality, drug addiction, education inequality, poverty, environmental degradation).

Apply the following three questions to the issue you want to evaluate:

Why do I see what I see?

What facts exist that contribute to the issue?

Why do things stay the same?

What are the barriers to change?

Why do things change?

What actions could be taken to initiate change?

Chapter Summary

1) Three factors contribute to continuous system functioning: 1) self-organization, 2) hierarchy, and 3) resilience.

2) Suboptimization is when a sub-element of a system dominates, diminishing overall system functionality and resiliency.

3) Identifying the flaws in social systems is imperative to resolving complex social issues.

3

NOW, LET'S TALK ABOUT SYSTEMS THINKING

WHY YOU NEED SYSTEMS THINKING TO ANALYZE AND SOLVE PROBLEMS

Aerial Shot by Henry Bauer

In a small fictional town named Lakeside, the central lake that had once been the community's pride began to suffer. Fish populations diminished, algae blooms became rampant, and the once-clear waters turned murky. The immediate reaction of the town council was to address the visible symptoms: they introduced chemicals to combat the algae and restocked the lake with fish. However, these solutions were short-lived. The algae returned, and the new fish died off.

A systems thinker named Morgan joined the council and suggested they look at the lake as a whole system, not just address the immediate problems. Morgan asked, "Why is the algae blooming?" and "What has changed in the lake's ecosystem?"

Upon investigation, they discovered that a new upstream factory in the town of Walen was discharging waste into the river. This waste contained nutrients that fed the algae. Additionally, they found that overfishing had removed many of the natural predators of the algae-eating species, causing an imbalance.

Using this system perspective to zoom out to the larger view, the town council negotiated with the factory to treat its waste. They also implemented fishing regulations to maintain ecological balance and introduced native aquatic plants to absorb excess nutrients.

Over time, the lake's health was restored. Fish populations stabilized, the water cleared, and the algae blooms ceased. Instead of treating individual symptoms, Lakeside's community addressed the root causes by understanding the system's interconnections. Morgan's systems thinking approach saved the lake and taught the town the importance of looking at the bigger picture.

The Wider Lens

Fantasy, Wall, Door by WiR_Pixs

Systems thinking replaces examining the parts with examining the whole. Analysis occurs by taking an object or concept apart to understand it. Systems thinking, on the other hand, puts an object or concept into the larger whole to understand it[1].

For example, if you want to examine drug addiction using systems thinking, you would zoom out to see where drug addiction is situated in the whole of society. Most importantly, in systems thinking, your focus will shift to look at the relationships and interconnections of societal elements and then examine the characteristics of the whole, rather than merely the individual parts.

A traditional analysis of drug addiction is that it is a personal problem for individuals connected to their health, willpower, and moral viewpoint. Treatment focuses on the individual.

When applying a systems thinking approach to drug addiction, the focus shifts to interrelated elements which include not only the individual psychology and physical characteristics of an addict. Also included in systems thinking are the relationships with family and friends, cultural norms, socioeconomic status, regional policies, drug availability, state and federal laws, and judicial processes.

Also, with a systems thinking approach, feedback loops are examined. What sensory information is received by a person's body and how does the body respond? For example, a person's body signals high stress levels and the response is turning to drugs for relief. Drug addiction causes further stress, becoming a suboptimization factor, with the addiction dominating.

Negative life events like family violence or poverty, as well as how drug use is criminalized, are other contributing factors[2]. When applying a systems thinking approach, it becomes evident that societal issues must be addressed at the same time individual issues are addressed.

Recently, the recognition of the impact of prescribed opioids on the drug addiction problem provided a leverage point—reducing opioid prescriptions (a small action) could lead to a large overall system improvement.

Systems thinking leads to a multi-stakeholder, collaborative approach to problem-solving. For the drug addiction issue, healthcare providers, the judicial system, the criminal justice system, drug manufacturers, communities, and families work together toward a holistic solution.

Drug addiction is a large, complex, social issue. The concept of "zooming out" can be applied to smaller systems. A local library that loans books to community members is relatively simple.

If we zoom out to see the local library as part of a larger whole, we see how it serves as the hub of a community network. The library, in addition to loaning books, promotes literacy and lifelong learning and is a community event space. The library hosts children's story time, authors' readings, book clubs, workshops, and political events.

The library provides internet and computers for local citizens to aid in homework, research, and job searches. The zoomed-out view shows the centrality of the library within the community. It has interconnections with the education, cultural, and employment systems. It is central to community engagement.

The Subtle Dance Of System Intent

The fundamental purpose of a system is not always obvious. Every subelement of a system has a purpose, but that is not the overall system's purpose[3].

Consider a hospital system. The purpose of the hospital system is to improve people's health and preserve life. The purpose of a single patient might be only to reduce pain superficially. The purpose of an individual doctor might be to advance to an executive role. The purpose of an administrator could be to balance the budget. You can see that these sub-goals could conflict with the overarching hospital system goal.

The successful functioning of a system relies on subelements aligning with the purpose of the larger system. This requires consistency in the relationships in the system.

For example, what if the relationship between doctors and hospital administrators changed so that the administrators reported to the doctors? Would patient care improve? The doctors would likely create a more patient-centric culture, but would their day-to-day administrative duties prevent them from enacting that culture? Doctors were not trained to be managers so may require additional training and education, taking away from their patient care time. What about administrative morale?

Systems thinking requires asking many questions to reveal the connections and interrelationships between all elements and to predict how changes in system elements will affect the larger system. All purposes, connections, elements, and interrelationships are essential for a system to function. However, changing the purpose of a system or the relationship between elements is likely to have the biggest influence on system behavior[4]. Let's look at an example.

Era's End: Fossil To Photon

Landscape, Mountain, Darling by ELG21

Changing the direction and purpose of a system is difficult. It is analogous to turning an aircraft carrier in the ocean quickly to move in another direction! We are witnessing such an attempt today as world leaders try to transform the energy sector from the current fossil fuel system to renewable energy sources. The purpose of the renewable energy system (wind, solar, and hydroelectric) is not only to provide energy but also to provide a sustainable, renewable source of power so that societies can continue to live beyond when fossil fuels are depleted. So, the purpose of our energy system has shifted.

Let's apply system thinking to this shift. What elements of the energy sector system serve as barriers to this transition?

Infrastructure - the current infrastructure was built to support fossil fuel production. New renewable energy infrastructure is expensive and uses different technologies.

Policies and regulations - new laws, regulations, and policies must be built to both incentivize and control renewable energies. At the same time, the new policies must act to disincentivize fossil fuel production and use. Policies, laws, and regulations are created in Congress which often results in political deadlocks.

Economics - many regions of the world employ millions of people in the fossil fuel industry. Jobs will be lost and significant training required to prepare workers for new jobs.

What are the factors that contribute to this change occurring?

Public awareness - People have become aware of climate change and the effects of fossil fuels on our environment. Society now demands improvement and cleaner energy sources, and they demand these new sources at a reasonable cost.

Technology - technological advancements continue in new renewable sources that make their implementation less expensive and more achievable.

I*nternational support* - many of the world's nations are working together toward renewable energy strategies. The Paris Agreement is an international collaboration that sets carbon emission reduction targets globally.

Using a systems thinking approach, we might ask, what small change to one element of the renewable energy system could be made that would have an impactful effect on the whole system?

Power Vaults: The Storage Enigma

One element of the renewable energy system is energy storage. Because wind and solar are not constant, energy must be stored to be available when wind and solar are not producing. This energy storage today occurs in lithium batteries, which are expensive, with prices increasing due to shortening supplies of both nickel and lithium. Batteries only provide several hours of storage due to the high cost[5].

A new report in Science Advances says the world's largest lithium stock might be in McDermitt Caldera on the Nevada/Oregon border, estimated to be 20-40 million metric tons. However, McDermitt Caldera is on indigenous tribal lands and the tribes are actively engaging in stopping mining[6].

The United States Department of Energy is developing a domestic supply chain for energy storage. Part of this plan is mandating the recycling of lithium batteries, which now occurs rarely[7]. This would be a small improvement that could have a significant effect on the consistent availability of stored energy.

Successful renewable energy use demands long-duration energy storage (LDES). LDES is defined as an energy storage technology or method that can discharge electricity at required power for no less than 8 hours and in a stable manner[8]. Now, renewable energy systems rely on gas generators to supply energy when solar and wind are not producing. This defeats the purpose of renewable energy as these generators negatively affect the environment through soil contamination and the release of greenhouse gases[9].

There are some interesting alternatives to battery storage, and the innovations are for long-duration energy storage. Goldman

Sachs invested $250 million in Hydrostor's compressed air storage. In this system, air is pushed into underground caverns (man-created) and stored under high pressure. When there is a need for electricity, a valve is opened, and the high-pressure air is used to spin electricity-producing turbines. Hydrostor has a large-scale project that can start as soon as 2026[10].

A second long-duration underground energy storage innovation is sand storage. Instead of storing compressed air in caverns, this method relies on lowering sand into unused mines. As the sand is lowered into the mines, the weight of the sand turns generators. Excess electricity then raises sand back to the top of the mine when renewable energy is plentiful. The self-discharge rate is zero so this energy storage method is ultra long term[11].

While the methods described above are attractive, they are geographically limited and must exist at a huge scale to be effective[12].

A modified use of sand for energy storage is demonstrated by Finland and is more realistic. In this system, hundreds of tons of sand are stored in large steel containers. The sand is heated to 500 degrees Celsius with renewable electricity and stored for use when renewable energy is unavailable. The goal is to make a storage container a thousand times larger. Similar energy storage can occur in water tanks[13].

Energy storage is one element of the larger renewable energy system where a shift from short-term to long-duration energy storage could catapult the implementation and stability of the larger system.

The Quiet Shift In Perspective

Conventional thinking is linear. A problem is identified, and the most immediate solution is applied. Conventional thinking identifies symptoms rather than underlying problems. Solutions appear obvious and are short-term. Long-term impacts are not considered, and because of this, unintended consequences emerge. Conventional thinking does not assume personal responsibility for a problem[14].

California faced challenges when it tried to switch from fossil fuels to renewable energy. They wanted to reduce carbon emissions because of environmental concerns. Their first idea was to use solar and wind energy. They asked questions about costs, technology, changes needed, and new rules. But they might not have thought about the bigger picture.

When using a broader approach, we need to think about more than just replacing one energy source with another. For instance, what happens when there's no sun or wind? How much energy can we store, and for how long? What if there's a gap in energy supply?

In 2020, California faced a problem. They didn't have enough stored energy when there was little sun or wind. This led to blackouts. People were asked to use less electricity, but it wasn't enough. The state had to cut power in many areas[15].

The government's plan was to have clean energy by 2030, zero-emission cars by 2035, and eco-friendly buildings by 2045[16]. But they were mainly focused on timelines and not the bigger picture.

We need to use a systems thinking approach to think differently. This means looking at everything as a connected system. Examine where feedback loops exist. We shouldn't just focus on the problem of using too much fossil fuel. We should consider how we produce, use, distribute, and store energy. We also need to consider how people's jobs and lives will be affected.

For example, storing energy is important. If we store more energy, it can become cheaper, and more people will want to use it. This can help avoid problems like the blackouts in California. But we also need to think about the effects of storing energy. Creating big storage solutions can harm the environment and use a lot of water.

In conclusion, when we try to solve a problem, we need to think about the whole system. Every action has a reaction, and we must be prepared for that.

Samsø Island, Denmark - Power Shift

Samsø, Denmark by Rotsee[17]

Samsø Island lies off the coast of the Jutland Peninsula in Denmark. It is a farming community, with a population of about 4000, which harvests the first potatoes every year. What makes this island community unique is that for over 20 years, 70% of its heating needs have come from highly controlled furnaces at district heating plants that cleanly burn straw, and all of the island's electricity is generated by huge wind turbines owned by the community[18].

Samsø Island's clean energy initiative began when it won a national competition to find a Danish Renewable Energy Island that would use renewable energy sources to become energy self-sufficient. The Danish Energy Authority funded the project. At first, the island citizens were not positive. They worried about the noise and how the wind turbines would look. The project leader, Hermansen thought that if the citizens owned the windmills, their resistance would diminish. He was right.

Everybody living on the island near the turbines could invest in them. As a community cooperative, they purchased 11 turbines, which created enough electricity for each of the island's villages to be energy self-sufficient. Before installing the turbines, environmental analysis considered how minimal impact on the wildlife and environment would be accomplished. Additionally, ten more offshore turbines were installed to offset car, tractor, and ferry emissions. Many citizens on Samsø Island use electric cars, powered with solar energy.

Danskt lantbruk och vindkraft på södra Samsø, fotograferat från Vesborg fyr by NewsØresund - Johan Wessman[19]

Future plans on Samsø Island include complete independence from fossil fuels by 2030. Some vehicles and tractors are still gas-powered. Recently, the island purchased a dual-fuel ferry that can run on liquified natural gas.

By 2007, only 10 years after beginning the project, Samsø Island produced more renewable source energy than it used and became a net exporter of renewable energy. Samsø Island's success is exemplary of community engagement, systems thinking, and holistic planning. Hawaii's Sustainable Molokai community and Hepburn Wind, a community-owned wind farm in Australia, were modeled after Samsø Island.

Conclusion

Complex social issues like drug addiction and poverty require "zooming out" to see the specific problem within the larger whole. These types of problems have typically been examined from an individual or family level. However, history shows that this approach has not solved these complex issues.

With the systems thinking approach, an individual's drug addiction is placed into the larger society as a whole, and every element influencing and contributing to the drug addiction is examined, as well as the societal elements that are affected by the drug-addicted individual.

Systems thinking contributes to finding holistic solutions to world issues like shifting from fossil fuels to renewable energy. Only with a systems thinking approach can the positive and negative implications of making this shift be understood. Systems thinking ensures that both short-term and long-term effects of major system changes are identified and considered.

Action Steps

The first step you can take in developing your system thinking is to practice "zooming out" to see the larger system within which a particular system exists.

Start small.

Examine a fish tank. There has to be a balance between food, water quality, and oxygenation. If there is a change in one element, like reduced oxygenation, other elements will be affected. When you zoom out on the fish tank, what other systems and subsystems affect the fish tank and the fish's health and well-being?

Think about a garden in your yard. What are the elements of the garden itself? When you zoom out, what other systems interconnect with that garden?

Your personal budget is a system. Identify the elements of that system. Zoom out and identify other systems that impact your personal budget.

Chapter Summary

1) Systems thinking puts an object or concept into the larger whole to understand it.

2) When applying systems thinking, you will "zoom out" to situate a scenario in the whole of society.

3) Systems thinking shifts the focus to interrelated elements.

4) Systems thinking examines the feedback loops in a system.

5) To successfully function, a system's subsystems and elements must align with the purpose of the larger system.

6) Changing the purpose of a system or the relationship between elements will have the biggest influence on system behavior.

4

SYSTEM SECRETS

THE ARCHETYPES THAT MAY TRAP US AND HOW TO AVOID GETTING CAUGHT IN THEM

The Parable Of Liora's Well

In the heart of the desert town of Liora stood a magnificent well, the town's primary water source. As the town's population grew, the well's water level began to drop, alarming the residents.

The town's council, eager to address the issue, decided to dig the well deeper. This "fix" seemed to work temporarily as the water level rose. Liora's residents rejoiced, believing their water troubles were over. However, as time passed, the water level began to drop even more rapidly than before. It turned out that the deeper well had tapped into a less replenished underground reservoir, leading to what is known as the "Fixes that Backfire" archetype, where a short-term solution leads to an unexpected consequence.

Desperate, the council sought the expertise of a renowned engineer, Alex. Alex introduced a modern pump system that drew water more efficiently. With this new system, the town's water supply seemed stable once again. However, this solution made the town dependent on the pump. Over time, as the pump required maintenance and occasional repairs, the town's skills in traditional water conservation methods faded, and their reliance on the pump grew. This was a classic example of the "Shifting the Burden" archetype: a short-term solution that shifted the burden to a new problem.

One summer, a sandstorm damaged the pump, rendering it inoperative. The town was left without its primary water source, and the people panicked. The forgotten traditional methods, which once sustained the town, had been overshadowed by their reliance on the pump.

A wise elder named Rohen stepped forward to help. He remembered the old ways of water conservation and rainwater harvesting. He taught the town to build cisterns, use water-saving techniques, and reintroduce native plants that required less water.

As the citizens of Liora rebuilt their relationship with water, they realized the traps they had fallen into. The quick fixes, while offering immediate relief, had long-term consequences that they hadn't foreseen.

The story of Liora's well became a lesson for future generations, illustrating the importance of understanding system archetypes and the traps that can ensnare even the best intentions.

Invisible Schematics: System Archetypes

In the intricate dance of systems, patterns emerge—recurring behaviors and structures that manifest within contexts, from biological ecosystems to corporate organizations. These patterns are called system archetypes, and they are foundational blueprints that can help us decode the often complex and intertwined relationships within systems. The archetypes help us recognize similar patterns, name them, and use the archetypes to leverage effective system changes. System traps are a subset of a system archetype illustrating counterproductive or difficult behavior patterns in a system[1].

System archetypes are not only theoretical constructs; they are powerful tools that provide insights into the recurring challenges and dynamics that systems face. By understanding these archetypes, we can anticipate potential pitfalls, design more effective interventions, and navigate our way toward desired outcomes.

Think back to the situation in California, when they rapidly moved to renewable energy without thinking about long-term implications. The outcome was rolling blackouts and citizens with significant periods of no electricity. It is easy to see the Fixes that Backfire trap at play in the California scenario.

Imagine trying to solve a maze without understanding its layout. You might find yourself trapped in loops or hitting dead ends. System archetypes offer a bird's-eye view of the maze, illuminating common traps and pathways to success. Whether it's "Limits to Success" that highlight the constraints systems might face or the "Shifting the Burden" that underscores the dangers of short-term fixes, each archetype tells a story—a narrative of system behaviors that have been observed time and again.

This exploration into system archetypes will not only introduce you to these fundamental patterns but will also equip you with the knowledge to apply them in real-world scenarios. As we delve deeper, you'll discover that these archetypes are not just abstract ideas but are, in fact, reflections of the world around us. They are the silent scripts that govern the behavior of systems, waiting to be understood and harnessed.

A system archetype is a repeating pattern of reinforcing and balancing feedback. Understanding the archetypes helps you quickly identify a system issue, find the root cause, and plan an effective intervention. Some system archetypes contain traps because they emphasize a pitfall that a system can encounter.

Fixes That Backfire

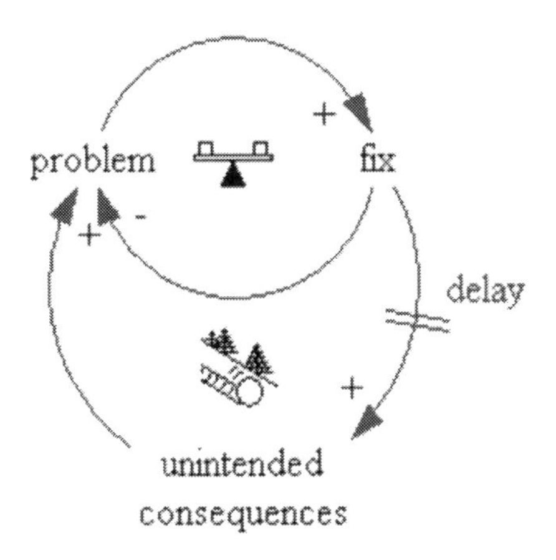

Causal Loop Diagram - System Archetype "Fixes That fail" by B Jana[2]

The Fixes that Backfire archetype usually appears as a long-term unexpected outcome after a short-term solution is put into place[3]. If we go back to our drug addiction scenario, taking painkillers after a serious injury provides excellent short-term relief. The long-term unexpected outcome may be addiction to the drug.

We previously discussed the Colony Collapse Disorder that resulted in significant reductions in bee colonies. In trying to determine the causes of this event, it was noted that a pesticide used to protect plants from pests unexpectedly affected the bees' nervous system and disoriented them so they could not return to their hives[4]. This was a fix to preserve crops with a highly unexpected result for bees that agriculturists could not have foreseen. It was a fix that backfired.

Another example of Fixes that Backfire archetype is if improved fishing nets are introduced to the market to increase catches. The unexpected result is overfishing which depletes the fish stocks.

Tragedy Of The Commons

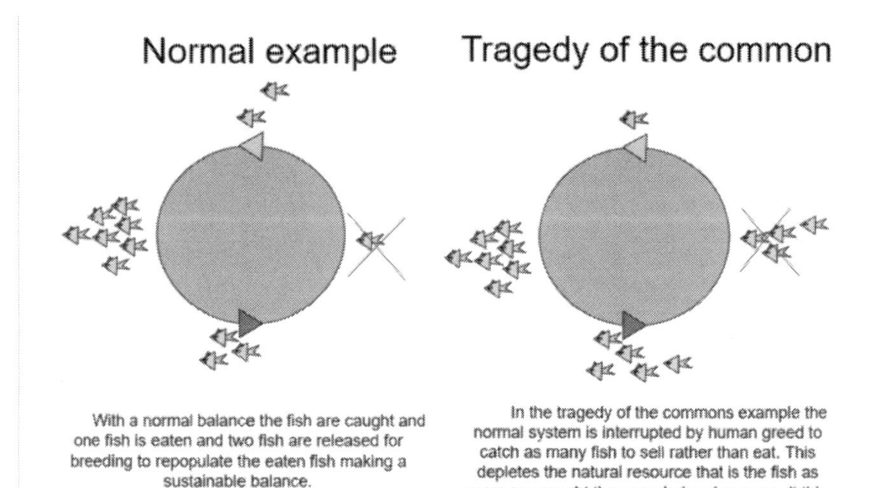

Causal Loop Diagram - System Archetype "Tragedy of the Commons" by B Jana[5]

The Tragedy of Commons archetype is present when a collective community resource is negatively affected by the actions of individuals[6].

Here is a fictional example of the Tragedy of Commons:

In a tiny coastal town named Seaside, there was a bountiful fishing area called Moon Bay. For generations, the Seaside residents relied on the bay's substantial marine life for their livelihood. Fish were abundant, and the fishers had a mutually understood rule: only catch what you need for the day. They wanted to keep the fish population healthy and sustainable.

One summer, word spread outside of Moon Bay, and commercial fishers from neighboring towns began to fish there. The outsiders did not know the cultures and traditions of the Seaside fishers. They wanted to bring in as many fish as possible, thinking of their immediate profit.

The Seaside fishers, instead of defending their commitment to a healthy and sustainable marine life, wanted to enjoy the same profits as the outsiders were experiencing. The Seaside fishers began competing with the outsiders.

Within only a few years, Moon Bay's rich marine life had dwindled drastically. The ecosystem had been disrupted and even the marine life that was not the target of the fishers declined. The bay, which had previously provided for all Seaside residents, was now barren.

If everyone had continued to take only what was needed from Moon Bay, it would have continued to flourish. Instead, greed resulted in depleting a shared resource.

On a very large scale, the global climate crisis can be seen as a Tragedy of Commons archetype. The entire Earth's atmosphere and climate system are shared resources for all nations and peoples. As large corporations in industrialized countries emit greenhouse gases into the atmosphere for corporate profit, environmental consequences were overlooked. Each country or individual company saw their emissions as minor and did not think of the pooled emissions.

Individual companies generate these emissions, but all people share in the negative environment. The common in this situation is the climate system's stability, which is being disrupted. It is necessary now for all nations to recognize their shared challenge of balancing individual country's interests with the collective good of a stable climate.

Limits To Success

What happens when an individual finds a miraculous diet that allows people to lose fifty pounds in four weeks? The success is weight loss, but the unintended result is that the diet results in mineral deficiencies and reduced health. Additionally, the limit is that the diet is not sustainable. The individual reverts to prior eating habits and gains the weight back.

The Limits to Success archetype is a pattern where growth cannot continue linearly forever. Limits will arise and should be planned for. Identify the limit obstacles early and eliminate them before they appear[7]. Initial success creates momentum that is not sustainable.

Think of a new streaming service that begins to offer a particular show that is very popular and the user base increases exponentially. Soon, the organization's servers cannot handle the volume, the streaming service goes down often and now users are dissatisfied and move to a more reliable service.

When the Limits to Success archetype is present, the approach must be to recognize and plan for the limits that are sure to occur. Similar to Fixes that Backfire, the short-term view looks good, but the long-term unexpected outcome has to be considered.

Shifting The Burden

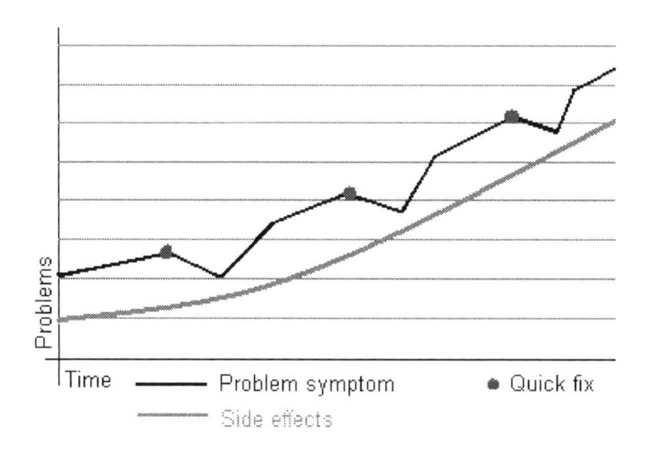

Graph to Represent the Fixes That Fail Archetypes Behavior by Svarnyp[8]

Think of the Wack-a-Mole game. You push the mole that pops up in one hole down, only to have another pop up in a different hole. This is similar to the Fixes that Backfire archetype. However, in Fixes that Backfire, short-term solutions have unintended consequences, while in Shifting the Burden, a quick fix that is implemented does not address the root cause and draws resources away from a more permanent solution[9].

Here is an example that will show the difference. A new drug is released to improve heart disease. However, side effects can damage the liver. This is a Shift the Burden archetype because a new problem emerged in a new organ in the body.

If that same new drug was released and later found not to have the intended effect of improving heart disease after long-term use, that is a Fix that Backfires archetype because short-term the solution worked, but there was an unintended consequence of the effect of the drug diminishing with long-term use.

Similar to Shifting the Burden is Shifting the Burden to the Intervenor. In this scenario, an external party (the intervenor) provides the symptomatic solution[10]. We saw this in the Parable of Liora's well when the city relied on the engineer who installed a pump to provide a more consistent water supply. The town then became dependent on the pump and the engineer who created it, losing their knowledge of the more natural ways to preserve and use water sustainably.

Drifting Goals

In the Drifting Goals archetype, standards and goals are gradually lowered due to short-term distractions or pressures. This can result in a significant performance reduction over time. When examining a situation in the Drifting Goals archetype, you will look at the following: a) the desired state, b) the current state, c) the existing gap, d) how to close the gap, and e) what pressures or distractions exist to lower the standard. Of key importance in this archetype is the importance of maintaining original, long-term standards when facing short-term challenges[11].

We see the Drifting Goals archetype often in education settings, where the pressure to improve graduation rates leads to a reduction in grading standards. The original goal was to strive for high academic standards to achieve quality education. The overall outcome is that there are more high school graduates, but they are poorly prepared to do well in college.

We also see this archetype in environmental regulations. The original goal is to set tight pollution limits to protect the environment and public health. However, pressure from the auto industry might lead to the government lowering the limits. The result is further degradation of the environment.

Escalation

When thinking of the Escalation archetype, think of an action that is perceived as a threat by another individual or organization. If actor A takes action to improve their own situation, actor B could feel threatened and act to improve their position as well. This competitive interaction can continue to repeat and escalate. What needs to be identified in this Escalation situation is a) what is the competitive element (e.g., price, prestige, political achievement), and b) who are the competing parties. Escalation leads to an inability to collaborate[12].

The arms race provides an excellent example of Escalation. We see this between countries that try to increase their military position, causing other countries to fear losing power, so they also increase their military position, and the arms buildup continues back and forth between the countries, making collaboration and cooperation impossible.

Social media breeds the Escalation archetype when two significant influencers disagree about an issue publicly, pulling in their own followers, so that the influencers and their fans engage in verbal warfare with each other in an escalating war of words.

Policy Resistance

When multiple players within a system push against a policy for different reasons, the result often ends in a neutralization of effects and none of the intended outcomes occur. This is the Policy Resistance Archetype. Constant competing goals intensify everyone's effort to put the system in the state they want it, but no one faction wins[13].

The introduction of the Affordable Care Act (ACA) is a good example. The goal of the ACA was to increase how many Americans obtained health insurance, while simultaneously reducing health care costs[14].

Many factions resisted the ACA. The republican party took steps to repeal and diminish the law. Some states resisted and would not offer the ACA, resulting in inconsistent healthcare coverage between states. Major insurance companies pulled out of the insurance market in states that adopted ACA, which reduced competition and raised premiums. The public was divided. As a result, the ACA's impact was significantly reduced.

Success To The Successful

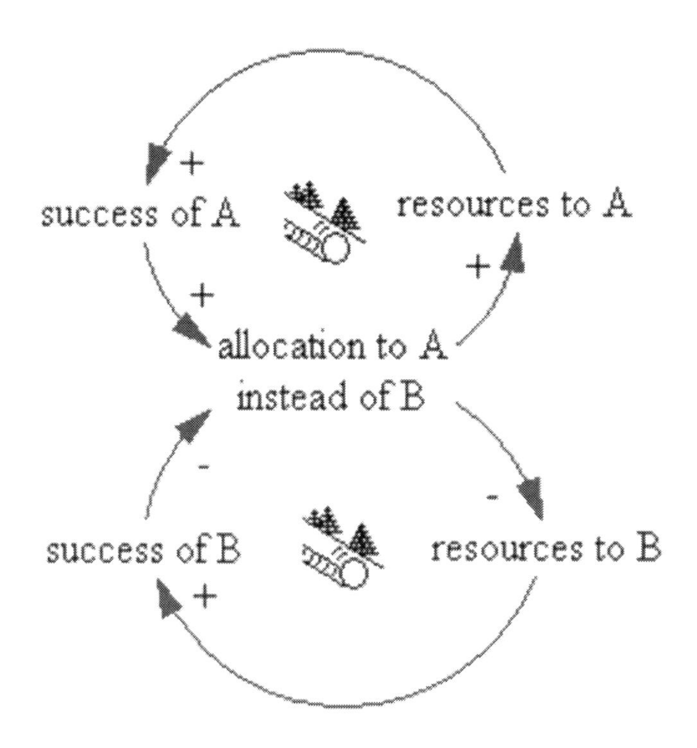

Causal Loop Diagram - System Archetype "Success to Successful" by B Jana[15]

When a person or organization uses their wealth, inside information, special privileges, or special knowledge to create more money, privilege, and access, we call this the Success to the Successful archetype[16]. It is a reinforcing feedback loop that quickly creates a two-class system of winners and losers. Athletes who win are highly compensated and, therefore, have the ability to purchase elevated training and advanced equipment to ensure they will keep being the winner.

Rule Beating

When deceptive action is taken to avoid the intent of a system's rule, the Rule Beating archetype is in play. Key to this archetype is that Rule Beating always gives the appearance that rules are being followed. Rule Beating can result in large distortions to a system.

The Volkswagen Emissions Scandal in 2015 is a good example. The company installed software that detected when their vehicles were being tested, and the software would then alter the car's performance to look as though the cars were complying with emission standards. When testing did not occur, the car's performance reverted to a lower emission standard, 40 times higher than permitted in the United States[17].

It was later found that many automobile manufacturers also altered test results. Volkswagen was fined and had to recall several vehicles. When rule beating occurs, the resulting damage extends beyond those that skirted the rules.

Seeking The Wrong Goal

When a goal does not authentically reflect the desired system outcome, the Seeking the Wrong Goal archetype is present. This archetype can be represented by using an incorrect metric. The goal might be focused short-term rather than long-term. A system might see improvement in the misaligned goal so there is reinforcement to continue on that path, and further lead away from the correct goal.

The opioid crisis began with a U.S. healthcare goal to manage patient pain more effectively. The American Pain Society referred to pain as a vital sign and hospital doctors were mandated to give effective pain control. To achieve this new mandate, the prescription of opioids led to the fastest results. Unfortunately, this over-prescribing led to addiction and increased the availability of opioids. Overdose deaths increased significantly[18].

Using System Archetypes

The system archetypes cannot be interpreted rigidly. More than one archetype might fit a certain situation, or you might combine archetypes to describe a system or scenario. The archetypes should serve to stimulate your thinking and help you put what you are seeing into a pattern that can be further analyzed. Looking at a situation from the perspective of two different archetypes might show different dimensions of the same scenario. The archetypes are a creative aid in your systems thinking.

Beneath The Flow: Unearthing Systemic Pitfalls

A system trap is a recurring pattern within a system that leads to counterproductive outcomes or persistent problems. Well-intentioned actions or interventions can lead to traps that result in unintended negative consequences. System traps highlight the complexities and intricacies of systems and underscore the importance of understanding the underlying structures and feedback loops that drive system behavior[19].

Decision-makers who do not understand system archetypes can fall into traps through uninformed or ineffective responses to challenges within the system archetype. Let's look at how traps emerge within different system archetypes and explore the way out of the trap.

Navigating The Shadows Of System Lures

Fixes that Backfire

In the Fixes that Backfire archetype, a trap emerges when a solution is focused on immediate symptoms rather than the underlying structure that causes the problem. Adopting a systems thinking approach, there are several strategies to navigate out of the trap.

1. Zoom out and imagine the unintended consequences that might occur. Think beyond the immediate problem and short-term solution. Apply long-term thinking and sustainable solutions.

2. Identify the root cause of the symptoms. Examine feedback loops that are driving the symptoms. Simulation and system modeling are useful tools to see how system elements interact in different contexts and to test different solutions before implementation.

3. When an intervention is implemented, monitor its results continually and adapt the solution based on the monitoring results[20].

An example is a city trying to relieve traffic congestion and building more roads to fix the immediate area of congestion. However, over time, more homes and businesses are built on these new roads, and long-term congestion is even more than the original. Using systems thinking to zoom out and take a long-term holistic view of the traffic issue results in looking at promoting carpooling, public transportation, and improved urban planning to reduce the need for long commutes.

Shifting the Burden

The Shifting the Burden trap is like the Fixes that Backfire in that symptoms rather than root causes have been addressed in the solution. Overreliance on this solution undermines efforts to address the underlying cause, and the root problem worsens.

1. Identify both a symptomatic and fundamental solution so that the root cause can be addressed.

2. Apply the 5-Whys method to see deeper into the problem to its root cause.

In the 5-Whys method, the question why is repeated five times[21].

a. Problem - the city is experiencing severe traffic congestion.

Why? There are too many vehicles on the road during rush hours.

b. Why? People drive their own cars instead of using public transportation.

c. Why? People think public transportation is slow and unreliable.

d. Why? The city has not updated its public transportation infrastructure in many years.

e. Why? City planning prioritized road development and car use in their strategic plan.

f. The root cause of traffic congestion is city planning prioritizing car use.

3. Collaborate so that diverse perspectives will more quickly reveal the root cause.

4. Do not overly rely on the symptomatic solution. Contain it to symptom relief while continuing to evaluate the root cause.

5. Examining feedback loops in the system can reveal if the symptomatic solution could be reinforcing the larger problem[22].

A perfect example of where a symptomatic solution reinforces the larger problem is substance abuse. A person who takes a drug to relieve stress creates a feedback loop where the more the drug is taken, the more dependent the person becomes, increasing the root problem of stress. To navigate out of this trap, a foundational solution must be sought through collaboration with a counselor and support group to identify the root cause of the stress. The symptomatic solution and foundational solution should be implemented simultaneously.

Tragedy of Commons

A situation where individual activity for self-interest exhausts a shared resource, leading to its degradation or demise, is called a Tragedy of Commons trap. A collective loss is the result. To navigate out of this trap, do the following:

1. Establish clear boundaries for the shared resource's use, defining limits on who can use it and to what extent.

2. Provide incentives for conserving the resource. Reward sustainable use and conservation investments[23].

3. Educate the community about the result of misuse of the shared resource and the collective benefits of sustainable use of the resource.

4. Monitor usage of the shared resource and implement regulations to limit the use[24].

Think of the community parks in your city. They are a shared resource for all residents who live in the city. However, parks can become overused, particularly when there is combined use by residents and migrating homeless people. Systems thinking can view the park as a system element within the entire city and set boundaries on when the parks are open, and the types of activities allowed. The city can designate certain parks or other areas for people experiencing homelessness to ensure they have safe shelter. Community-sponsored clean-up groups can be incentivized.

Drifting Goals

When standards are lowered in response to short-term pressures, the Drifting Goals trap arises and typically results in a long-term performance reduction. The following systems thinking approaches can be applied to navigate out of this trap.

1. When pressure arises to lower standards, reaffirm core values and long-term vision and make decisions that support those values.

2. Frequently review goals to see if they have drifted from the core values. If they have, realign them.

3. Keep all stakeholders involved in supporting the core values.

4. Put continuous feedback loops in place to identify quickly when a core value is drifting.

5. Always maintain a long-term view when faced with short-term pressure. This will avoid a compromising attitude.

6. Do not change goals in response to pressure or challenges. Explore the root cause of the immediate problem[25].

We set many personal goals in our lives, and this is an area where the Drifting Goals trap can be seen often. You might set a weight loss goal, and when it is not met, you reduce it (for example, you change the goal from losing 10 points in a month to only losing 5 pounds). Assuming you set a reasonable and healthy goal, instead of revising it, revisit and modify the diet plan. Continue modifying diet plans until the one that achieves the desired goal is found.

Success to the Successful

The Success to the Successful trap is often called the rich get richer. This trap results in reduced diversity and monopolies. To escape this trap, the following approaches can be taken.

1. Robin Hood's "steal from the rich, give to the poor" approach is a form of wealth redistribution. Increasing taxes to the rich is a form of wealth redistribution, as is providing scholarships to students of low socioeconomic status rather than scholarships that only reward academic excellence.

2. Ensure new entrants into a system can compete fairly. This can relate to all systems (education, sports, socioeconomic)[26].

3. Encourage collaboration as well as competition.

4. Ensure that multiple paths to goals and achievement exist.

5. Reduce positive feedback loops that allow success to feed on itself by introducing balancing loops.

6. Redefine success and ensure multiple paths to success[27].

AT&T grew into a dominant monopoly in the telephone service market in the mid-20th century. The organization's monetary success allowed it to invest in research, infrastructure, and technology that cemented its dominance. It became difficult for competitors to enter or stay in the telecommunications industry.

Luckily, the United States has antitrust laws, and the Department of Justice broke up the monopoly in 1974. AT&T was broken into multiple companies called Baby Bells. The Baby Bells managed local telephone service, and the original AT&T managed long-distance service and research[28].

The balancing loop in this system was the creation of the multiple independent Baby Bells, making it so no single entity could dominate the market. Competition returned to the market, with consumers having choices, lower prices, and improved service.

Shift the Burden to the Intervenor

In the Parable of Liora's Well, we saw the Shift the Burden to the Intervenor trap. Realizing that the first solution of deepening the well did not improve the water supply as they expected, the town leaders turned to an engineer who designed and installed a pump to improve the consistency of the water supply. They shifted the burden to the intervenor, the engineer, and then became dependent on the well that the engineer installed. The trap is the dependence on the intervenor. As we saw with the Liora residents, they lost knowledge of sustainable preservation and use of water the more they relied on the pump. To escape from the trap, the following approaches can be taken.

1. Sometimes the intervenor is not obvious, so in those cases, the intervenor must be identified.

2. Identify the circumstances that caused the system to turn to the intervenor for a solution.

3. Even though an intervenor is brought into the situation, do not abandon old approaches and knowledge. Run parallel solutions that address the issue.

4. Monitor the system's dependency on the intervenor and correct when the dependency is too high.

5. Collaborate with the intervenor to continue building additional solutions that incorporate a diversity of knowledge and understanding[29].

Social media, like Facebook, provides an interesting example. A primary goal of Facebook was to engage users for longer periods. To accomplish this, the company turned to an intervenor, sophisticated algorithms, which contributed to prioritizing emotional and controversial content which kept people using Facebook and for longer periods.

The Facebook system became dependent on the algorithms to continually increase user engagement and Facebook use. The trap was that users were ultimately only shown content that supported their existing beliefs. Polarization between people increased, as did misinformation dissemination. Diverse viewpoints presented to users disappeared.

To navigate out of this trap, Facebook and other social media platforms gave users control over content that came through their feeds. The organizations had to make their algorithms transparent to the public. Third-party fact-checkers were hired to prevent misinformation. Third-party screen time trackers emerged to help users know how much time they were on social media. While many of these algorithms are still used, the intervenor has been balanced with other tools and technologies[30].

Rule Beating

Often when Rule Beating is discovered, a response is to further strengthen the rule to destroy a self-organizing rebellion[31], but this typically leads deeper into the trap. To navigate away from this trap, the following steps can be taken.

1. Evaluate the existing rules to determine if they lead to the desired outcome. Rules that are too narrow can be unfair. Some rules are counterproductive.

2. Add supporting rules and metrics so that multiple system elements work toward the same goal.

3. Analyze the factors contributing to the problem. What feedback loops are reinforcing the behavior?

4. Implement audits to ensure rules are complied with, but also make measurement transparent so that breaking the rule is more visible.

5. Create a value-respecting culture that understands the need for the rules and is committed to supporting them.

6. Gather feedback regularly on the satisfaction and dissatisfaction with the rules.

7. Discover whether rule-breaking is somehow being incentivized.

8. Create significant consequences for anyone gaming the system[32].

Numerous athletes fall into the Rule Beating trap when they take performance-enhancing drugs. This has led the World Anti-Doping Agency (WADA) to make three revisions to their World Anti-Doping Code, between 2006 and 2017. The Code synthesizes anti-doping policies, regulations, and rules within sporting organizations to provide consistency. The revisions have continuously resulted in stronger, more robust rules and testing standards[33]. What seems to be missing is educating the athletes on the value of fair competition to change the culture of the sports teams to a value-focused culture.

Seeking the Wrong Goal

To move away from the Seeking the Wrong Goal trap, it is necessary to recognize that the intended outcomes are not being achieved. It is important to begin by reevaluating the goals to ensure they align with the desired outcome.

The metrics used to measure the goal's achievements are relegated to the goals. If the goals are incorrect, the metrics will be as well.

1. Evaluate what the original purpose of the system was.

2. Determine if the stated goals align with that purpose.

3. Ask if the metrics used to measure the goals accurately reflect the state of the goals.

4. As needed, change the goals and/or metrics. You might have the right goal but the wrong metrics to measure achievement of the goal[34].

We can see a movement away from Seeking the Wrong Goal in city planning. Many cities prioritized car travel, building a complex network of roads and freeways that prompted much building along these transportation routes. The goal was to increase access to housing and businesses by car travel. The result was increased pollution and non-walkable cities.

City planning is shifting toward decreased car use by supporting effective public transportation, walking cities, and cultural and park spaces throughout the city. The new and improved goal is to create cities that improve the quality of life for its residents.

Invisible Scaffolds: The Unseen Forces Shaping Our Reality

System archetypes represent a visible, recurring pattern within a system. However, when an individual sees those patterns, they are interpreted based on that person's experiences and the culture they were raised in. This interpretation is the mental model a person uses to evaluate what they observe[35].

Mental models emerge from knowledge we already have and allow us to synthesize new knowledge to further our understanding[36].

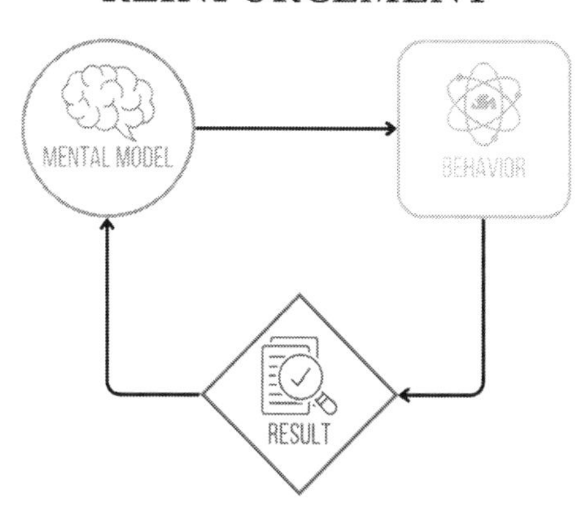

Mental Models Diagram by Wisdom University

Mental models shape our perceptions and actions. They interpret our observations, often filtering them through our personal experiences. Viewing the world through tinted glasses colors our reality, just as our cultural upbringing might equate material wealth with success.

Such models can lead us to make choices based on these skewed perceptions, like prioritizing high-paying jobs over passion. System archetypes reveal how our mental models can overlook complexities, leading to unintended outcomes. For example, believing in unlimited growth in a finite system can result in system failures.

Similarly, certain behaviors stem from deeply ingrained models: a distrustful individual might act defiantly, a child from a violent home might be perpetually fearful, and an illiterate person might rely solely on verbal communication. Athletes, having been praised from a young age, might seek constant validation.

Ultimately, our mental models are interpretative lenses—some empower, others limit. Embracing change and new perspectives fosters adaptability and resilience in our ever-evolving world.

In the world of systems thinking, where interventions are implemented, understanding the mental models of all people involved can help anticipate and respond to implementation challenges[37].

Conclusion

The story of Liora's well showed us how system archetypes help us recognize similar patterns, name them, and use the archetypes to leverage effective system improvements. Each system archetype has a trap that reveals challenging behavior patterns in a system. Understanding the archetypes helps us recognize a system problem and provide a solution more quickly.

It is equally important to understand our personal mental models that might be influencing how we interpret the system archetypes. In the next chapter, we explore systems thinking models that offer frameworks from which to evaluate a complex system.

Action Steps

Identify a relatively small system you are familiar with. This could be your child's school, the grocery store where you shop, the organization you work for, or even city parks.

Consider a problem that exists in that system.

Review the system archetypes that we have discussed and select one or two that you feel describe the system in its current state.

- How does the archetype describe the system?
- What is the trap in the archetype?
- How can you navigate away from the trap?
- Consider your own mental model that might have influenced how you interpreted the system.

Chapter Summary

1) Systems have reoccurring patterns and behaviors called system archetypes.

2) System archetypes describe a system and help us predict how that system will behave in the future.

3) Each system archetype has a trap that results in the system acting in a counterproductive manner.

4) Past research regarding system archetypes and traps has identified methods to navigate away from the trap.

5) Mental models can interfere with how people interpret a system archetype.

6) Reflecting on our mental models is necessary to become more aware of how they influence our thinking.

5

NAVIGATING THE COMPLEXITIES OF THE BIG PICTURE

THE KEY SYSTEMS THINKING MODELS AND HOW YOU CAN INTERPRET, COMMUNICATE, AND DEEPEN YOUR UNDERSTANDING OF THEM

Wolves in a Forest by patrice schoefolt

The Wolves Of Yellowstone: Unexpected Engineers

In the 1920s, wolves living in Yellowstone National Park had been hunted to extinction. This had an unexpected consequence on Yellowstone's ecosystem. The elk population boomed, and they overgrazed young aspen and willow trees. The domino effect saw beaver numbers decline, as these trees were critical for their sustenance and dam construction. The subsequent erosion of streams disrupted habitats for otters, ducks, and fish.

In 1995, in an attempt to repair this ecosystem, U.S. wildlife officials reintroduced 41 wolves into Yellowstone, and the result was transformative. The elk population dropped significantly, and the ecosystem was restored over time[1].

As the elk stopped overgrazing, vegetation flourished, and the beaver population rebounded. The cascading benefits included stabilized riverbanks and enriched habitats for otters, ducks, and fish.

This is a vivid testament to the intricate interplay of life, where every element, regardless of how small, plays a crucial role in the balance of the system. The reintroduction of wolves, initially seen as just predators, showcased their indispensable role in maintaining ecological harmony.

It is hard to make sense of such complex interrelationships. How can we predict the ripple effect that a single change will make on an extensive system? Enter the mental models of systems thinking.

Decoding The Matrix: Systems Frameworks

Human beings have always sought ways to make sense of the intricate web of reality, leading to the development of various mental models. These cognitive tools simplify complex phenomena, allowing us to predict outcomes, make decisions, and understand the world.

Just as we have personal mental models (as discussed in Chapter 4) that guide our individual perceptions and behaviors, there are also more universal frameworks that help us decipher larger, interconnected systems. Enter the realm of systems thinking.

This discipline introduces models like stocks, flows, and feedback loops, offering a structured approach to understanding system component's dynamic interplay. By shifting from a narrow, isolated perspective to a holistic view, systems thinking provides tools to navigate and influence the complexities of our interconnected world.

In systems thinking, we examine complex problems as part of a larger system. Building on our previous discussions, we delve deeper into the foundational models of systems thinking:

Feedback loops are cycles where a system's output influences its input and can either amplify (positive) or stabilize (negative) systems.

Stocks and flows were introduced in Chapter 1. Stocks are intangible or tangible elements that can change over time, while flows represent the rate of this change.

Delays or bottlenecks represent the time lag between an action and its subsequent impact on a system.

System boundaries, as discussed in Chapter 1, define the limits and scope of a system. Boundaries might change as systems are altered.

Leverage points are strategic places in a system where a small adjustment might lead to significant outcomes.

Emergence is a key underlying foundation of systems, entailing that the system's collective output is greater than the mere sum of its individual elements[23].

We can examine complex systems by employing these models, predicting repercussions, and devising efficient solutions. In this chapter, we focus on three core system thinking models: stocks, flows, and feedback loops.

Silent Anchors: Stocks In Focus

A stock, whether physical, like water in a reservoir, or abstract, like an organization's goodwill, represents the memory of a system's progressing flows[4]. Consider tree rings: Each ring indicates not only the tree's age but also reflects the environmental conditions of specific years. The rings reflect drought, abundant rain, and disease[5]. The tree records its historical life in a natural archive.

Annual Rings by Couleur

Whispered Currents: The Power Of Flows

Flows dictate the movement in and out of stocks. For instance, water flows from rivers to bays and then to oceans. Flows encompass various actions between birth and death or deposit to withdrawal. Reflecting on the Yellowstone example, the interplay between stocks (elk, wolves, vegetation) and flows (grazing, predation, growth) becomes evident.

The Dance Of Stocks And Flows

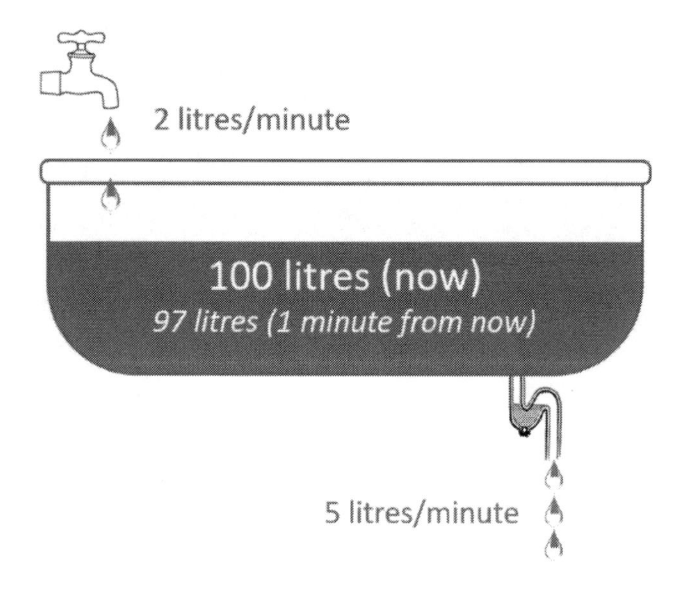

How the Rate Water Flows Into and Out of a Bathtub Changes the Quantity or "Stock" of Water Each Minute by Kim D Warren

Stocks and flows are the heartbeat and pulse of systems, forming the framework for how they operate and develop. A bathtub is a simple representation of the interrelated stock and flows. The water in the bathtub is the stock, while the faucet and drain represent the flow. The water level inside the tub changes based on the inflows and outflows controlled by the faucet and drain.

Invisible Dialogues: Feedback In Action

Feedback loops are recurrent processes where a stock's status influences a flow, which in turn affects the stock[6]. Consider a home fish tank: The water's purity (stock) determines the filter's operation (flow), ensuring a conducive environment for the fish. As impurities in the water increase, the filter's operation also increases, bringing the water back to purity.

Did you ever play with a slinky, making it walk down your stairs?

A Person Playing a Slinky Spring Toy by Tara Winstead

It is an example of a feedback loop in action. As the slinky is poised on the top stair, with part of it hanging over the edge of the step, when it is released, gravity pulls the hanging part of the slinky down toward the next step. As the front of the slinky moves downward, it creates a tension that pulls the rest of the slinky forward. The tension causes the back part of the slinky to catch up and move to the next step[7].

The movement of one part of the slinky influences the next part, and the action/reaction cycle repeats. The descending slinky is a series of interconnected movements, a feedback loop, that moves the slinky down and forward.

Hidden Tides: The Invisible Drivers Of Action

Sea Water and Seashore by Ashish Sonawane

Systems have both internal and external forces. The external forces are the most visible system elements. Our attention is drawn to these elements because they are obvious. However, it is the internal infrastructure that is often invisible that controls a system's behavior[8].

Think about your smartphone. When you hear about a new release, you focus on the cool, visible features that have been added: screen size, battery life, camera quality, and new display options. However, there are essential hidden forces behind what is visible. Software developers, hardware vendors, and market researchers were vital in bringing the new features forward. A behind-the-scenes supply chain brings the elements together for assembly and delivery to the stores. The successful launch that you see as a customer cannot happen without these invisible forces.

Think of your favorite local restaurant. The visible forces you experience are the staff, the taste of your favorite dish, and the ambiance. Customer reviews are about these external factors. Behind the scenes is a flurry of activity that ensures the experience you have. Inventory management ensures fresh ingredients are always in stock. Staff training programs and chef expertise are required. Customer feedback is reviewed and acted on. As the customer, you do not see these forces in action, but they are critical for maintaining the restaurant's reputation.

Periods of economic downturn illustrate the internal and external forces. Often, when high unemployment and high inflation exist in the United States, the external forces that are visible are often the President and their political party, who receive the blame. The quick fix is to vote in a new leader to solve the problem. The hidden forces include the capitalistic economic system's volatile structure, market regulations, and consumer behavior. Any attempt to improve the economy must include a holistic understanding of how the visible and hidden elements interact.

Systems In Practice: The Healthcare Paradigm

The multifaceted U.S. healthcare system epitomizes the substance of systems thinking. Tangible elements like hospitals, doctors, nurses, patients, and equipment are stocks, while staff training, patient admissions and discharges, and medical supply purchases depict flows.

However, examining only stocks and flows does not reflect a wide-angle view of the healthcare system. We must also identify the hidden elements, like administrative structure, supply chains, and regulatory agencies like the Centers for Disease Control and Prevention (CDC). which played a pivotal function in driving behaviors during the COVID-19 pandemic.

The underlying relationships between the hidden and visible elements reveal the system's complexity. An examination of the healthcare system's response to the pandemic, from dwindling masks, personal protective equipment (PPE), and glove supplies to the progression of vaccine availability and the reduction of COVID-related deaths, underscores the importance of the deep analysis demanded.

While the public witnessed the healthcare system's overt actions, the CDC's research, guidelines, and protocols were the hidden forces guiding the visible behaviors.

Amplification And Balance: Loop Dynamics Unveiled

Balance

The easiest way to understand the dynamics of feedback loops is to look at simple examples in your everyday environment. Think of a cold day when you set your heating system thermostat to 75°F. The purpose of the system was to monitor the room's temperature and compare it to the desired setting, adjusting the heater on and off to maintain the target temperature.

The room's temperature is the stock. The thermostat has rules it must follow. The heater turns on when the room temperature is below the desired setting. When the room reaches the desired setting, the heater turns off. It is a system of continuous monitoring and adjusting.

The feedback loop in the heating system is a negative feedback loop that works to stabilize a system by adjusting to any deviations from a target point. This feedback loop responds by opposing a change, adjusting, and bringing the system back to its desired state. The system maintains equilibrium[9].

Another example is the classic predator-prey relationship, which is a negative feedback loop in nature. We can look at lions and zebras in the African Savanna.

Lion and Zebras, Etosha National Park, Namibia by Frank Vassen[10]

1. Initial condition: Zebras are plentiful, and lions have plenty of food, so the lion population increases.

2. As the lion population grows, they eat more zebras, decreasing the zebra population.

3. Now, there are fewer zebras for the lion's food supply, so the lion population declines by starvation or migration to a new area.

4. When the lion population declines, the zebra's predators decrease, so the zebra population increases.

5. The cycle repeats, keeping this ecosystem in equilibrium.

6. The negative (balancing) feedback loop ensures a stable system, but the system is also more resistant to change.

Amplification

The second type of feedback loop is a positive-reinforcing loop. In this type of loop, a change in a system variable leads to an increase in that same variable, which then increases or amplifies the initial change[11]. Positive, in this sense, means the direction of the change, not that it is necessarily "good."

One of the simplest examples of the positive-reinforcing loop is what we call the snowball effect. A snowball, rolling down a hill, collects more and more snow. As the snowball's surface increases, it must collect even more snow, so it grows at an expanding rate.

Here is another example of a positive-reinforcing feedback loop. When she was 25, Anna started a savings account that yields a 5% compounding interest rate. Her initial deposit was $5,000, and she deposited $5,000 every year until she turned 65.

At the end of year 1, Anna has her initial deposit of $5,000 + her earned interest of $250. Her balance is $5,250. She deposited another $5,000 at the beginning of year 2, so now she has $10,250. At the end of year tw0, she has $10,250 plus her earned compound interest of $512.50 for a balance of $10,762.50. When she reaches age 65, she has deposited $200,000 over 40 years. However, her account balance is

$600,000 due to the positive-reinforcing loop of compound interest.

Anna's case illustrates a desirable result, but positive-reinforcing loops can also result in damaging outcomes. Think about credit card debt.

1, Initial action: John uses his credit card to make a purchase because he does not have enough cash.

2. The result: John cannot pay off the balance at the end of the month.

3. Consequence: Interest is charged on the remaining balance.

4. Feedback loop: The added interest increases the total debt, making it even more difficult to pay off.

5. Reinforcement: As the debt grows, John uses other credit cards to pay daily expenses because he has less available cash, leading to more interest and debt.

In these examples, an initial action leads to a sequence of events that intensifies the original action. The positive-reinforcing feedback loop, when left alone, results in devastating system collapse or exponential growth.

Visual Alchemy: Decoding Through Diagrams

Now that you have learned about the mental models most used in systems thinking, it is time to discuss the diagrams and charts often used in systems thinking. Charts and diagrams allow you to see a complex system at a glance. They also help you describe a system to others with visuals that are easier to understand.

Diagrams more clearly depict the relationships between system elements than can be explained in words. Loops are easy to identify. Changes can be made to them, and results can be simulated before changing the actual system. They also document the current state of a system. Charts and diagrams foster a holistic perspective.

Here is a stock and flow chart depicting a product manufacturing and sales system with a negative feedback loop that monitors the desired stock level as new items are produced and sold. The stock could represent any product (smartphones, furniture, or cars). When the stock falls below the desired stock level, production increases.

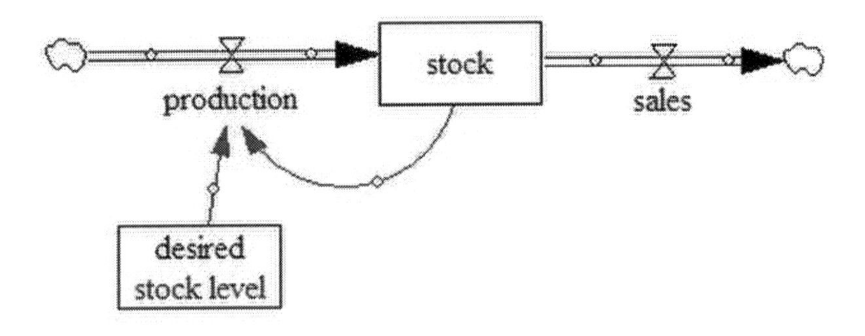

Stock and Flow Chart[12]

Let's look at a stock and flow diagram that represents the zebra/lion (predator/prey) scenario that we discussed. You will see that added factors of zebra and lion fertility, birth rate, and mortality rates are included.

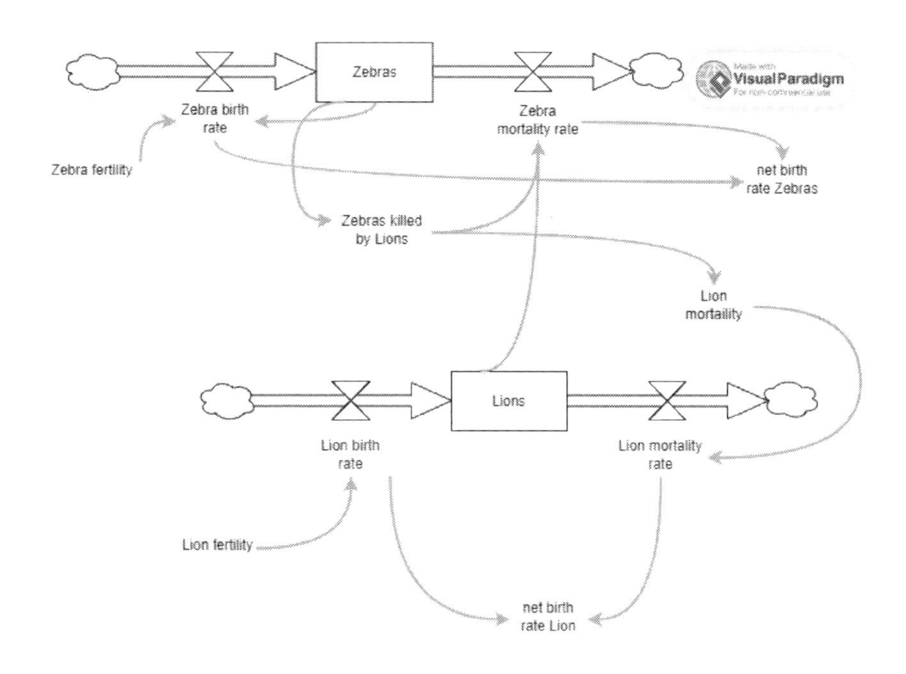

Lion and Zebra Stock and Flow Chart by Susan Ferebee

We know in the predator/prey loop that a negative feedback loop keeps the ratio of lions and zebras at an equilibrium, where neither becomes dominant. If the zebras have a high birth rate in one year, there will be more food for the lions, which will reduce the lion's mortality rate. However, as more zebras are eaten by lions, their mortality rate increases, which reduces the lion's food supply, decreasing their population. The stock and flow diagram clearly illustrates the negative feedback loop between predator and prey.

The predator/prey cycle can also be shown in a graph that depicts the population of each over time. The graph clearly shows that the increase in one coincides with the decrease in the other, and the rise and fall of predator and prey populations is consistent over time in a repeating cycle.

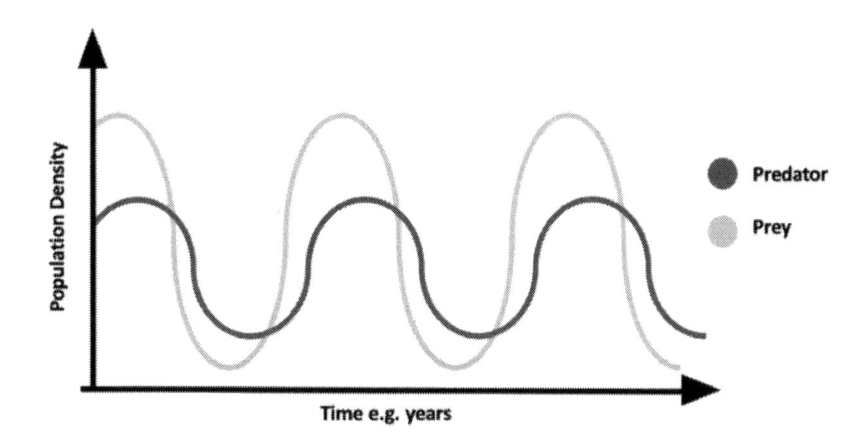

A Generalised Graph of a Predator-Prey Population Density Cycle by Hczarn[13]

Showing a positive reinforcing effect is often the easiest to see in a chart. For example, the concept of a continuous and amplifying increase through compound interest is readily illustrated in this chart:

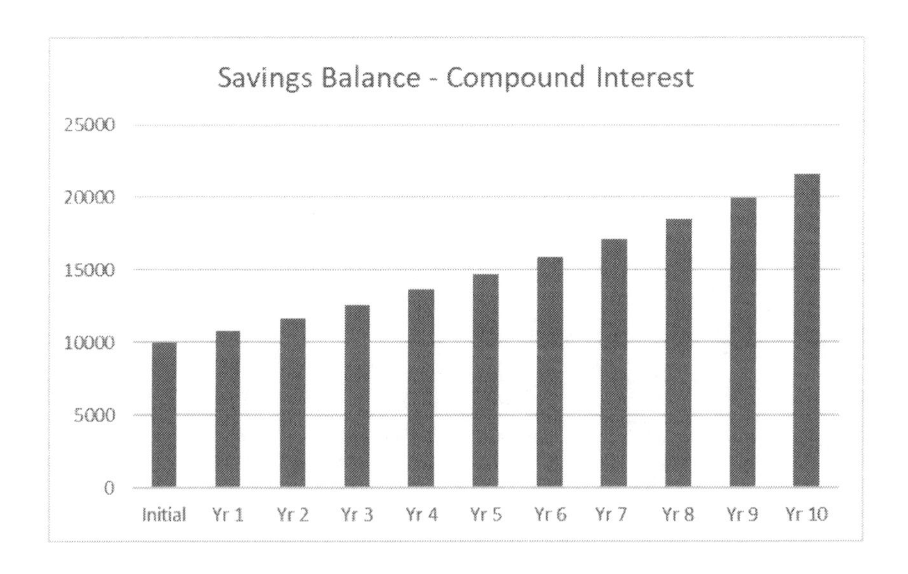

The Silent Harmony Of Complexity

Complex systems are composed of numerous interconnected components or agents. These components interact with each other, often in non-linear ways, leading to emergent behaviors that cannot be predicted by examining individual parts in isolation. Examples of complex systems include ecosystems, economies, human bodies, and even social networks.

We have learned that complex systems are often robust and adaptable. But what contributes to a high-functioning complex system? There are three important factors:

1. Redundancy - having multiple system elements that can perform the same function makes a system resilient[14]. An example would be in our predator/prey scenario. In most ecosystems, if one predator is removed, another might continue controlling a particular prey. In Yellowstone, for example, when wolves were removed, coyotes became the predator in the region, but they were not as effective as wolves. When the wolves were returned to Yellowstone, they not only preyed on elk but also on the coyotes.

2. Feedback loops - positive-reinforcing and negative, as discussed in detail in this chapter. Feedback loops allow a system to endure consistently over time.

3. Diversity - having diverse components ensures a system can handle a wide range of challenges[15]. A good example of system diversity is seen in polyculture farming, where farmers plant multiple crops together in the same field. Central American farmers, for example, plant beans, squash, and corn

together in a shared field. The beans provide nitrogen to the soil, while the cornstalks serve as bean poles. The farmers plant low-growing squash between rows because the leaves shade the ground, conserving moisture and inhibiting weed growth.

Companion Planting by Anna Juchnowicz[16]

Additionally, mixing crops inhibits the spread of certain pests, and if one crop fails for any reason, the other two might survive, ensuring food security[17].

A city is an example of a complex system that includes housing, utility services, waste management, transportation, healthcare, and more. Using system models to depict the system allows urban planners to predict how a change in one area might affect others.

New York City, for example, transformed the unused elevated railway track called the High Line into an urban linear park between 2009 and 2014. Changes in other city elements occurred as a result:

1. The housing property values in the surrounding areas increased significantly, leading to the construction of luxury apartments. This led to a shift in the demographics.

2. Local businesses also flourished in the area, improving the city's economy.

3. The park fostered walking and reduced traffic congestion in the surrounding neighborhoods.

4. Previous residents faced rent increases, and many were displaced, leading to gentrification.

5. Ecologists must amplify the visibility of these conversations. Tackling its intricacies demands fresh perspectives, novel mathematical approaches, and substantial, unwavering financial support[18].

Conclusion

Understanding complex systems through system models provides invaluable insights into their behavior, resilience, and adaptability. By recognizing the underlying patterns and structures, we can better navigate, manage, and optimize these systems for a desired outcome. Whether preserving an ecosystem, designing a resilient urban infrastructure, or improving healthcare outcomes, system models are indispensable tools in our quest to comprehend and harness the power of complexity.

System diagrams allow us to visualize complexity and to see a large system as a whole. A visual depiction helps us quickly see interrelated elements and helps ensure that when considering a change to one element, we will examine how that change might affect other elements. They aid comprehension and facilitate communication, strategic planning, and problem-solving within any discipline.

In the next chapter, we will delve deeper into the concepts of resilience, self-organization, and hierarchy that we introduced in Chapter 2. Additionally, we will learn about leverage points for system behavior change.

Action Steps

Examine a small system in your home (heating, cooking, water, sound systems, or alarm systems).

Use the standard stock and flow symbols (below) to draw a stock and flow diagram that describes the system.

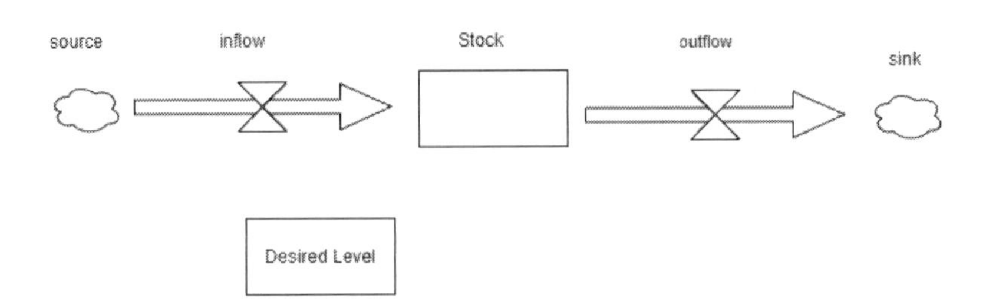

Questions to ask as you begin your diagram include:

1. What are the main reservoirs or accumulations in the system that represent quantities that change over time? These will be your stocks.

2. How do resources or information enter or leave the stock? Is there a process that increases or decreases the stock? These will be your flows.

3. What is the source of the input? Where do they eventually leave the system? These are your sources and sinks.

4. Are there cycles where changes in stock affect flows, and that, in turn, affects the stocks again? These are your feedback loops.

5. How are the stocks and flows related to each other?

6. What is the boundary of the system? This determines what you include in your diagram.

Chapter Summary

1) The three key models used in evaluating a system are stocks, flows, and feedback loops (negative and positive reinforcing).

2) Stocks represent the memory of a system's progressing flows. Stocks are physical or abstract concepts that can be measured like a store's inventory.

3) Flows dictate the movement in and out of stocks, like newly manufactured products flowing into a store and then flowing out as consumers purchase them.

4) A negative feedback loop strives to maintain system stability by adjusting to deviations from a target point. An example is a thermostat.

5) A positive-reinforcing feedback loop occurs when a change in a system variable increases that variable, amplifying the initial change. An example is compounding interest.

6) Charts and diagrams allow you to see a system holistically. They enhance your ability to analyze systems, make predictions about the system, and test potential changes within a system to evaluate how a change might affect other parts of the system.

7) Three factors contribute to a high-functioning complex system: a) redundancy, b) feedback loops, and c) diversity.

6

BENEATH THE SYSTEMS BLUEPRINT

UNCOVER THE COVERT CHOREOGRAPHY OF SYSTEMS TO KNOW WHERE AND HOW TO PIVOT

How do the hidden attributes of a system silently shape its outcomes, and how can we harness these attributes? Every system, whether an ecosystem, a city, or a business, operates based on a set of underlying attributes that are sometimes not noticed but act as the foundation of the system's function. Once we understand these attributes, we are better able to optimize the system.

Do you wonder how seemingly minor tweaks in a system can lead to profound transformations? What if you could pinpoint where to apply the slightest pressure and yield the most significant change?

Let's dive in and unravel the mysteries of system attributes and leverage points so we can steer systems toward desired futures.

The Hidden Architecture Of Sustainability

Sustainability Energy Tree by Gerd Altmann

In this chapter, we will take an in-depth look at resiliency, self-organization, and hierarchy—pillars of sustainable systems introduced earlier. Remember the resilient tree? It's time to explore resilience in more intricate systems.

Veiled Vigor: The Unseen Backbone

You might remember our discussion of the resilient tree and its use of redundancy, robustness, modularity, buffer capacity, feedback loops, diversity, subsystems, interconnection, communication between subsystems, and learning from the past to remain sustainable. A resilient system absorbs environmental disturbances, adapts, and continues functioning. Now, we will look at what resilience looks like in more complex systems.

In each of the following examples, we see the sheer robustness and diversity of the systems contributing to their resilience.

But they also have unique adaptations that led to their resilience.

The Silent Strength Of The Barrier's Bloom

Great Barrier Reef Corals by Gökhan Tolun[1]

The Great Barrier Reef off the Australian coast is the world's largest coral reef system. It covers over 2,300 kilometers and includes thousands of individual reefs. It is one of the seven natural wonders of the world.

When sea temperatures rise significantly, coral bleaching occurs. The Great Barrier Reef experienced bleaching in 1998, 2002, 2016, and 2017. In spite of these occurrences, the reef recovered remarkably due to several resiliency factors.

The Great Barrier Reef's survival was due to the diversity provided by over 400 coral species. While some species would be severely threatened by a specific threat, others were resistant and continued to flourish. The diversity was an important factor in recovering from bleaching as certain species adapted more effectively.

Interdependent relationships with other systems also support resilience. The parrotfish, for example, graze on algae, keeping the algae from smothering corals. This was particularly important during bleaching because the corals were weaker.

Part of the Great Barrier Reef's system is governmental and conservation organizations that work to strengthen and protect the reef. These groups grow coral in nurseries to transplant into damaged sections. The Great Barrier Reef Marine Park Authority serves as a feedback loop in the reef system by monitoring the reef's health for early detection of threats and timely interventions[2].

The Sundarbans' Silent Stand

Mangroves, Sundarbans, Forest by sarangib

The Sundarbans Mangrove Forest, stretching across Bangladesh and India, stands as the world's largest mangrove expanse and is the realm of the majestic Bengal Tiger.

Covering over 10,000 square kilometers, this forest is a testament to nature's resilience[3].

The Sundarbans is not just a forest but a vast mosaic of islands intricately woven together by a complex network of water channels. This unique land, one of the world's most expansive deltas, has been sculpted by the sediments of three mighty rivers: the Ganges, Brahmaputra, and Meghna, and further shaped by the ebb and flow of tides.

The Sundarbans is not only resilient in its own right but also bolsters the resilience of its surrounding environment. The intricate root systems of its mangrove trees act as natural shields against coastal soil erosion. Some mangrove species have evolved to excrete salt through their leaves, while others, through specialized roots known as pneumatophores, draw oxygen even during high tides, reaching deep beneath the mud[4].

This forest is a sanctuary for biodiversity, housing a plethora of species ranging from the Bengal tiger, Indian python, and saltwater crocodiles to 260 bird species, 334 plant varieties, and diverse deer species. The intricate web of interdependence among these species ensures a balanced and thriving ecosystem.

Beyond its immediate boundaries, the Sundarbans plays a pivotal role in climate balance, acting as a potent carbon sink. It sequesters carbon dioxide at rates surpassing even tropical rainforests, underscoring its global ecological significance[5].

Drawing parallels with the Great Barrier Reef, the Sundarbans' sustainability is a collaborative effort. Both India and Bangladesh have designated protected zones within this forest, imposing restrictions on activities like fishing and logging to safeguard this ecological treasure.

The Unscripted Dance Of Self-Organization

At the heart of many complex systems lies a phenomenon where patterns and order spontaneously emerge, not from external forces but from the system's own internal processes. This is the essence of self-organization.

It is a system's innate ability to adapt, restructure, and function in response to external challenges, all with the primary goal of survival[6].

Key ingredients fueling self-organization include:

- Interactions between system elements
- Feedback loops that guide adjustments
- Adaptation to changing circumstances
- Emergence of new attributes
- Protection of critical elements[7]

Such systems are the epitome of resilience and sustainability. They efficiently allocate resources, adapt to challenges, and offer diverse solutions, ensuring they are not easily toppled by a single point of failure. Moreover, they are hotbeds of innovation, where different elements experiment with solutions, and successful ones get amplified.

Let's examine how self-organization occurs in the highly complex system of the Internet.

The Internet: A Digital Dance Of Self-Organization

The Internet, a global network of interconnected computers and servers, is the epitome of a highly complex self-organizing system. The Internet began from a U.S. Department of Defense project in the 1960s, connecting only a few institutions.

Born from a U.S. Department of Defense project in the 1960s, the Internet has grown from connecting a handful of institutions to becoming a global digital behemoth. Today, it is a vast socio-technical tapestry, connecting billions and serving as a dynamic reservoir of human knowledge[8].

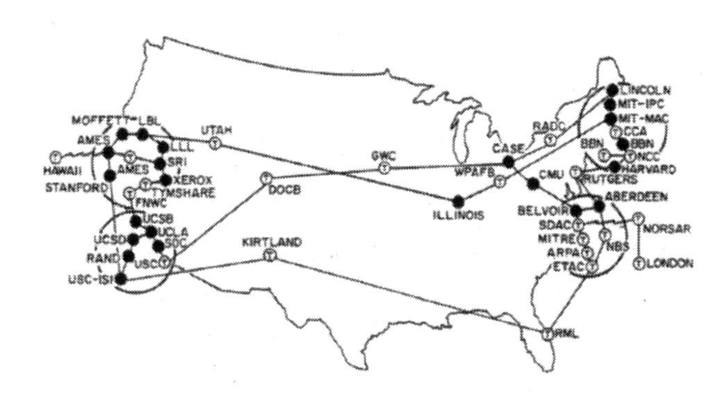

Arpanet Points in the 70s by Semaforo GMS [9]

Every day, this digital repository grows richer as we contribute content, engage in conversations, and tap into its knowledge. The beauty of the Internet lies not just in its vastness but also in its decentralized nature. It is a world without a central governing body, where countless independent networks coexist and collaborate.

The Internet, as a self-organizing system, relies on ground rules of standardized protocols that ensure every piece of data finds its way to the correct destination, while often zigzagging across many networks.

The various elements of the Internet (independent artists and content creators) ensure many independent inputs and choices into the system, enhancing the system's sustainability. This decentralized, self-organizing nature brings numerous benefits:

Unyielding strength - Like a living organism, the Internet constantly evolves. New devices are instantly integrated, and traveling data finds a detour, ensuring that it operates continuously. Its decentralized design is its buffer. If some parts fail, the collective remains robust.

Nurturing innovation - The Internet embraces creativity. Social media, streaming video and audio, blogs, and websites offer a creative playground for endless discovery.

Boundless expansion - Diversity multiplies as the Internet effortlessly scales to millions of new users and devices, all thanks to its self-organizing magic.

The Global Financial Market: A Web Of Interconnected Dependencies

The global financial market is a huge, intricate system that is complex and demonstrates self-organization. It comprises a multitude of stock exchanges, banks, investment organizations, and individual human investors, operating across all time zones, cultures, and regulatory environments.

In spite of the chaotic look and feel of this expansive financial market, the system is in continual self-organization. It balances endless individual decisions with overarching regulations. It adapts to a continuous stream of new real-time information, evolving to meet the changing needs of the global economy[10].

Some elements that contribute to the self-organizing behavior are:

Price determination - Stock, bond, commodity, and currency prices are determined through the collective actions of millions of traders. These actions lead to price patterns in the market.

Capital flow - Money moves fluidly across borders seeking the highest returns. Capital flows into promising economies and flows out of risky economies.

Information dissemination - Data and news are generated and consumed in the financial markets, leading to emergent market behaviors.

Risk management - The financial market system constantly adapts to new data to remain stable.

Regulatory feedback - Monitors the behaviors and stability of the market, intervenes when anomalies occur, and injects capital to stabilize the system.

Innovation and adaptation - A key attribute in a self-organizing system, is seen through new trading strategies, currencies, and technologies that emerge in response to evolving challenges and needs.

The dynamic interplay of forces makes the global financial market one of the most fascinating complex systems in the modern world.

The Hidden Harmony Of Sustainable Ascendancy

In the intricate world of systems, hierarchy plays a vital role in ensuring sustainability. Hierarchy, in this context, is the arrangement of elements and subelements within larger systems, forming a layered structure with each level operating at a different scale and with its own unique functions[11].

The tiered structure ensures efficient resource allocation, role specialization, and adaptability to changing circumstances. Hierarchies also reduce redundant actions and promote system resilience. According to Simon[12], the hierarchy provides functional simplicity and a way to manage inherent complexity.

The Amazon Rainforest: A hierarchical symphony

Amazon Rainforest by CactiStaccingCrane[13]

The Amazon Rainforest in South America spans 3-3.2 million square miles of which 80% is forested. Two-thirds of the Amazon is in Brazil. The Amazon River, in the Amazon

Rainforest, is the largest river in the world. During the high-water season, the river's mouth grows to 300 miles wide.

The Amazon moves water as it flows down the Amazon River, but the trees of the Amazon push vast amounts of water vapor into the atmosphere daily, which later falls as rain or is carried by airflow to other parts of South America. The movement is sometimes called "flying rivers[14]."

Being so vast, it is difficult to imagine a hierarchical structure underlying the Amazon Rainforest system, but it is a living testament to the power of system hierarchy in supporting sustainability. Let's examine the high-level elements in this hierarchy:

Micro level: Species interaction - Individual organisms like plants, animals, fungi, algae, and microorganisms each have their niche where they live and each plays a specific role in the ecosystem. The roles could be predator, pollinator, prey, or decomposer.

Mid level: Community dynamics - Individual species form communities based on needs and interactions. A particular tree might provide food and shelter to several bird species. Both cooperative and competitive relationships exist.

Macro level: Ecosystem interplay - Multiple communities combine to form unique ecosystems within the rainforest. Some examples are riverine, upland, and wetland ecosystems. Each supports different species and faces unique challenges.

Top level: The rainforest as a whole - The ecosystems combine to form the whole Amazon Rainforest as a single entity. High-level challenges like deforestation, climate regulation, and water cycle maintenance emerge[15].

Myrmecophyte Trees And Ants

Now let's look at a very small example in the vast Amazon Rainforest where symbiotic interaction between a tree species and ants illustrates a hierarchical relationship. Let's examine the hierarchical levels that exist:

The Forest Canopy - At the top level, the Amazon Rainforest has a distinct canopy layer, the upper layer of the forest that receives the most sunlight. This layer comprises tall trees and emergent species.

Myrmecophyte Trees - At the macro level, within this canopy, there are unique trees called myrmecophytes. They have evolved special structures called domatia, which are hollowed-out areas or swellings in the trunks and branches. The Myrmecophyte trees are at the macro level, forming an ecosystem

Ant Colonies - The myrmecophyte's domatia serve as homes for certain ant species, forming a mid level community serving the needs of both the trees and ants. In return for the shelter, the ants serve as a defense mechanism for the tree, attacking herbivores that try to eat the trees' leaves. They also keep other plants from growing at the tree's base.

The micro level emerges with the fungi and multiple soil microorganisms. The sustainability of the trees and the ants supports the sustainability of other surrounding ecosystems and the rainforest itself.

The Unseen Toggles Of Leverage Points

Leverage points are places in a system where a minor change results in a significant shift in a system's behavior. Complex systems are often counterintuitive; therefore, identifying the leverage points is also not intuitive.

Donella Meadows[16], author of the 2008 book *Thinking in Systems*, lists the following potential leverage points in a system in increasing order of effectiveness:

- Physical structure (stocks and flows)
- System parameters (numbers)
- Buffers (like the water reservoir of a dam)
- Delays
- Balancing feedback loops
- Positive reinforcing feedback loops
- Information flows
- System rules
- System self-organization
- System goals
- Cultural paradigms - Society's shared ideas

According to Meadows, creating a small change in a cultural paradigm or changing a system's goal are the most effective leverage points in a system. Rutherford[17] discusses the value of collective efforts in contributing to identifying and using system leverage points.

Often system interventions related to sustainability, target highly visible, but weak, leverage points - meaning the intervention won't result in transformation. To reach transformative leverage, three areas should be examined: a) reconnecting people to nature, b) restructuring institutions, and c) reexamining how knowledge is generated[18].

Let's look at an example of a successful leverage point applied in healthcare systems.

Handwashing: A Simple Turn Of Health

Wash Your Hands by Amousey

A small and inexpensive change in healthcare procedures, implementing handwashing requirements, led to significant improvements in health outcomes. Both the World Bank and the Center for Disease Control and Prevention (CDC) have highlighted the cost-effectiveness of hand hygiene promotion, labeling it as the most impactful health-related action to combat disease[19].

The Global Handwashing Partnership[20], in 2017, noted that a mere investment of $3.35 in handwashing campaigns can achieve health improvements comparable to spending thousands on immunizations. For example, Unicef, in their global handwashing campaign, informs people of vast inequalities in access to handwashing at home, school, and healthcare facilities. They then promote Global Handwashing Day, which is supported by governments, international organizations, corporations, civil service groups, and schools[21].

Within hospitals, the prevalence of healthcare-associated infections (HCAI) has been a costly concern, with expenses reaching $6.5 billion in the U.S. in 2004. Yet, these costs can be substantially reduced by simply enhancing handwashing compliance among healthcare professionals.

In the intricate, vast maze of healthcare practices, handwashing surfaced as a pivotal leverage point. Just as a minor tweak in a machine can lead to outsized improvements in its performance, the simple act of handwashing became a linchpin in the healthcare system.

Its introduction and emphasis not only led to direct benefits in cleanliness but also set off a cascade of positive effects. Lower infection rates resulted in fewer treatments, shorter hospital stays, and reduced demand on medical resources.

There was a ripple effect that influenced hospital budgets, staffing needs, and general public health. Handwashing, a seemingly modest act, highlighted the sheer impact a single leverage point can have in transforming a complex system.

Looking at Meadows' list of leverage points, we can see that this intervention focused on information flows, disseminating information on the benefit of handwashing throughout the healthcare system. In hospitals, in addition to the information dissemination, rules were changed regarding handwashing compliance. And, at the most general level and in conjunction with the information flow, there was an effect on the cultural paradigm.

Unforeseen consequences and new leverage points

As with many large-scale behavioral shifts, shortcomings and unforeseen consequences also occur. With regard to hand hygiene, we see several:

1. An increase in alcohol poisoning in children under age 12 occurred in early 2020 due to increased exposure to alcohol hand sanitizers.

2. Excessive use of antimicrobial soaps and sanitizers can lead to antibiotic resistance[22].

3. Exacerbation of existing skin conditions through the use of alcohol-based sanitizers[23].

A leverage point to tackle these unforeseen consequences might be the promotion of integrated health and hygiene education. In an emergency response to COVID-19, a narrow, short-term solution was sought. Now, in response, public campaigns need a more holistic approach, focusing on all crucial aspects of hand hygiene, combined with environmental responsibility and overall health considerations.

Operation Green Bin

General Recycling Symbol by Team445

Remember when we did not have recycle bins and all our waste went into one garbage can? Landfill employees were tasked with separating recyclables from garbage. In 1980, Woodbury, New Jersey was the first city in the U.S. to order the public to recycle. The city provided curb-side pickup for recycled goods, but the system was not easy for consumers, who had to recycle different items into multiple small bins and carry them all to the street[24]. The following issues stimulated the need to improve recycling processes and attitudes:

Environmental concerns - Plastic waste polluted the ocean and endangered marine life.

Resource scarcity - Recycling aluminum requires 95% less energy than to produce from raw materials[25].

Public awareness and demand - As the public became more aware of marine life being damaged, by plastic straws as an example, they began to demand more careful use of and disposal of plastic products.

Legislation and policy - Many states began to mandate recycling. The U.S. Environmental Protection Agency (EPA), through the Resource Conservation and Recovery Act, began regulating hazardous wastes and landfills, and setting recycling goals[26].

Economic implications - Landfill costs were high.

Technological advances - There were significant advances in recycling technology[27].

A leverage point was needed. What small change in the waste management system could significantly impact increasing recycling? In 1995, in California, single-stream recycling (placing glass, plastic, and paper in a single large container) was introduced[28]. This single container was collected in the same way as normal garbage, adding little complexity to residents' lives.

This became a critical leverage point in the larger waste management system, aiming to shift behavior to more sustainable practices.

Referring again to Meadows' list of leverage points, this one small change in the waste management system contributed to changing a cultural paradigm to one where recycling became an everyday part of life by simplifying a process.

Unforeseen consequences and new leverage points

As in our handwashing scenario, some issues and unforeseen results accompanied the recycling behavior shift.

1. Inconsistent guidelines were provided as to what constituted recyclable plastic, paper, and glass. In particular, not all paper and plastic were recyclable and residents were confused[29].

2. Single-stream recycling increased participation, but it also resulted in increased contamination compared to multi-stream recycling where recyclables are separated by type[30].

3. Recycling is more environmentally friendly than producing new materials, however, increased recycling increases energy use and produced emissions[31].

4. Overemphasis on recycling overshadowed the importance of reuse and reduction, the other elements of waste management. Reuse and reduction are the most effective elements[32].

Again, we need to find the leverage points to address these unforeseen consequences. Standardized recycling guidelines could be a small leverage with a significant reduction in contamination resulting from incorrect recycling.

Increased promotion of reduction and reuse, in conjunction with improved recycling, provides a holistic approach to waste management promotion and education. Additionally, to better handle increased recycling behavior, recycling technology needs to be modernized.

Conclusion

As we close this chapter on the covert choreography of systems and their pivots, it's evident that beneath the visible facade of every system lies a complex web of interactions, hierarchies, and leverage points. These hidden attributes, often overlooked, are the true drivers of system behavior and outcomes.

From the resilient Great Barrier Reef and the majestic Sundarbans to the digital dance of the Internet and the intricate world of global finance, we've seen how systems, both natural and man-made, are shaped by their underlying attributes. These attributes, when understood and harnessed, can be the key to steering systems towards desired futures.

The concept of leverage points, in particular, offers a powerful lens through which we can view and influence systems. As demonstrated by the examples of handwashing in healthcare and single-stream recycling, even seemingly minor interventions can lead to profound transformations. However, as with all interventions, it's crucial to be aware of potential unforeseen consequences and be prepared to adapt accordingly.

In essence, the journey through this chapter underscores the importance of a holistic understanding of systems. It's not just about recognizing the visible components but delving deeper to uncover the hidden dynamics at play. By doing so, we equip ourselves with the knowledge and tools to optimize systems, ensuring their sustainability and resilience in the face of ever-evolving challenges.

As we move forward, let's carry with us the insights gleaned from this chapter, always seeking to understand beneath the blueprint of the systems we encounter. In doing so, we position ourselves to make impactful changes, steering systems toward a brighter, more sustainable future.

Action Steps

Think about a social system that you are familiar with. What is the primary goal of this social system?

What are behaviors that need to change within this system to reach that desired goal?

Think about the leverage points that exist in this social system.

Generate a set of ideas that represents a small shift in the system that might result in a significant change.

Chapter Summary

1) Sustainability requires resiliency, self-organization, and hierarchy.

2) System resiliency is increased with:

- Redundancy

- Robustness
- Modularity
- Buffer capacity
- Feedback loops
- Diversity
- Subsystem interconnection
- Communication between subsystems.

3) Self-organization is the system's innate ability to adapt, restructure, and function in response to external challenges.

Ingredients supporting self-organization are:

- Interactions between system elements.
- Feedback loops that guide adjustments.
- Adaptation to changing circumstances.
- Emergence of new attributes.
- Protection of critical system elements.

4) Leverage Points are places in a system where a minor change results in a significant shift in a system's behavior. Potential leverage points in a system are:

- Physical structure (stocks and flows)
- System parameters (numbers)
- Buffers (like the water reservoir of a dam)
- Delays
- Balancing feedback loops
- Positive reinforcing feedback loops
- Information flows
- System rules
- System self-organization
- System goals

- Cultural paradigms - Society's shared ideas

5) Every change to a system, even resulting in positive results, can and likely will have unforeseen consequences that need to be evaluated, looking for new leverage points to resolve the issues.

7

BREAK THE MOLD

EXPLORE SYSTEMS THINKING'S ROLE IN TRANSFORMING OUR WORLD

Web, Spider web, Network by fietzfotos

Deep within a dense forest, young Aria often ventured out on her own, exploring the wonders of nature. One foggy morning, she stumbled upon a splendid spider web, glittering with dew. The intricate design was enthralling, with each thread connected to another, forming a complex pattern.

Aria observed that a gentle breeze caused a single leaf to fall onto one side of the web. The entire web quivered to her amazement, even the parts far from where the almost weightless leaf landed. A tiny insect trapped on the opposite side of the net was jolted by the leaf's impact.

Curious, Aria gently tugged on a single thread on the web's edge. Again, the entire structure vibrated, affecting every part of the web and its inhabitants.

An old forest ranger, Mr. Elliot, who had been watching Aria's amazement, approached her. "It's fascinating, isn't it? How a single touch can affect the entire web," he remarked.

Aria nodded. "It's like everything is connected."

"Exactly," Mr. Eliott replied. "Just like this web, everything in our world is interconnected. A change in one area can resonate through an entire system. That's why we must be mindful of our actions, as they can have unintended consequences in places we might not even consider."

Aria left the forest that day with a new recognition of nature's intricate balance and connections that bind us all. She realized that, like the spider web, our society is a complex system where every action can have a ripple effect, no matter how small.

This chapter explores the complex web of systems thinking and its profound effect on understanding the delicate balance of social dynamics.

Rooted Realities And Silent Depths

Society's problems are complex and interwoven, with visible and invisible parts, all interacting. Because of this, solving the issues requires the following approach.

1. Avoid quick-fix solutions that undermine long-term effectiveness.

2. Set pragmatic expectations.

3. Plan for the long-term first, then create supporting short-term goals[1].

4. Aim for short-term successes that intentionally support long-term results and provide individuals with real hope rather than false promises.

To successfully address critical social issues like unemployment, criminal justice, climate change, education inequality, and ecology shifts, the ability to perceive and understand dynamic interconnections inside and across interrelated systems is paramount[2].

In thinking about social systems, David Peter Stroh, who authored the 2015 book *Systems Thinking for Social Change*[3], defines systems thinking as a means to accomplish a desired purpose. There is often a difference between the purpose a social system is currently achieving and the desired purpose.

Think about the U.S. prison system. The primary purpose is usually stated as being the rehabilitation of criminals and ensuring the public's safety. It is believed that rehabilitation will lead to reduced reoffending after release, and community safety will improve.

However, the United States has the highest incarceration rate in the world, but many who are released recidivate. The public is demanding criminal justice reform, but to do this successfully, there must be accurate facts and an understanding of the big picture view of the criminal justice system[4].

Think about the sub-systems that are part of the U.S. criminal justice system.

1. Prisons (federal, state, local, private, and tribal)

2. Other correctional facilities (jails, juvenile correctional facilities, immigration detention centers, military prisons, state psychiatric hospitals, and U.S. territory prisons)

3. Bail systems

4. Courts (federal, state, municipal, and juvenile)

The prisons and courts are stocks in our criminal justice system. The churn of suspects, convicts, and released prisoners through the courts and prisons represent the flows.

As individuals are convicted and incarcerated, there is a wide ripple effect throughout the families of the prisoners and crime victims as well as to the larger community.

Questions to ask to see the bigger picture include:

1. Why are so many people in prison?

2. Do prisoners provide free labor to private organizations? Do non-violent offenders belong in the same prisons as violent offenders?

3. Do they belong in jail at all?

4. Which prisoners are the most likely to be rearrested?

5. What is the root cause of most crimes?

6. How do crime victims perceive justice?

7. What does a thriving rehabilitation environment look like?[5]

Let's look at three real-world examples that reveal different systemic issues in the criminal justice system and examine them with a systems-thinking view. Then we will look at a successful prison reform that relied on long-term solutions.

Shadows Of Rikers

Rikers Island Jail by Tim Rodenberg[6]

In 2010, Kalief Browder, an African American teen from the Bronx, New York, was arrested for allegedly stealing a backpack. There was minimal evidence, and Kalief's story changed multiple times. Kalief was already on probation.

Despite inadequate evidence, the state decided to prosecute. And because Kalief or his family could not post bail, he was sent to Riker's Island jail to await his trial. He was in the island jail for three years without ever being convicted of a crime.

While in Riker's, Kalief was placed in solitary confinement for almost two of the three years and was repeatedly physically abused by guards and inmates. Court dates were endlessly delayed, and the case was finally dismissed.

However, Kalief had severe trauma from his experience in Riker's. He attempted suicide several times while incarcerated and succeeded in taking his own life in 2015, two years after his release[7].

Clearly, this scenario did not reflect the purpose of rehabilitation and protecting society from danger. The impact of Kalief's family's low socio-economic status is a significant element leading to his prolonged incarceration only because he could not make bail.

In this system, there appears to be a lack of needed feedback loops that monitor court delays to ensure that a non-convicted prisoner is not left in prison indefinitely. There was also no feedback loop that identified a non-convicted prisoner remaining too long in the jail.

Was there a leverage point in this system where a small change could have resolved this problem leading to a significant change in this criminal justice system?

Cycle's Silent Grip

In many urban regions across the U.S., local communities and law enforcement have seen a pattern where individuals arrested, imprisoned, and released are rearrested again within a short period. This points to a failure in rehabilitation programs or understanding the root cause of crime.

A 2021 report by the U.S. Bureau of Justice showed that about 62% of state prisoners released in 2012 in 35 states, were rearrested within three years and 71% rearrested within five years. Almost half of prisoners released in 2012, returned to prison within five years for violating probation or parole or being sentenced for a new crime[8].

As a systems thinker, you could diagram this arrest/rearrest cycle in a stock and flow diagram.

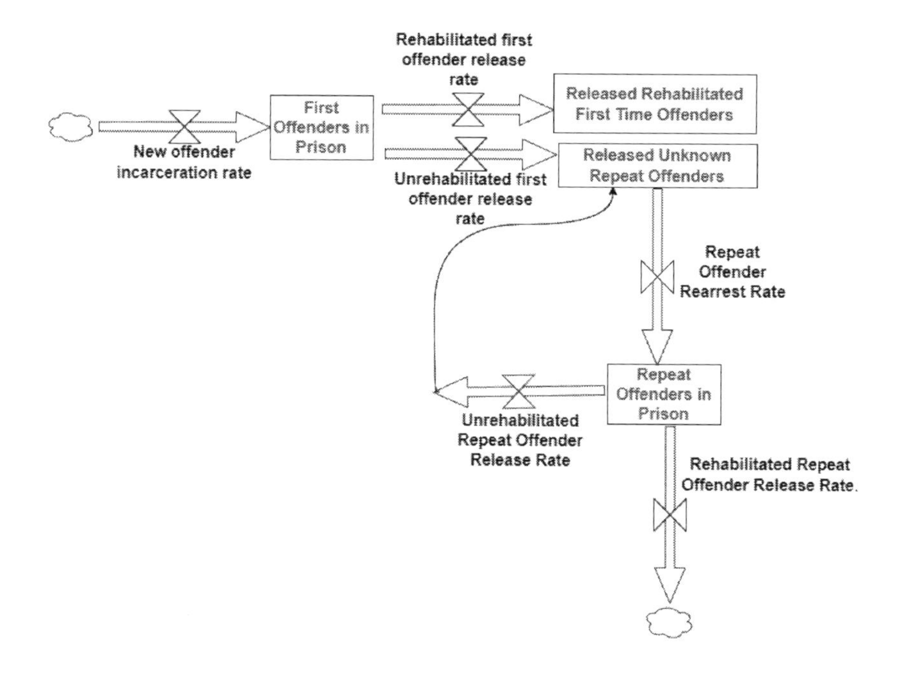

Arrest/Rearrest Cycle Diagram by Susan Ferebee

Statistics show that rehabilitation often does not work[9]. As a systems thinker, we then have to zoom out and think about what the wider system comprises. As we look through a wider angle than just the incarceration system, we would add the following:

Community and family environment - Returning released convicts to the same environment they came from puts them around the same influences that supported their criminal activity.

Socio-economic factors - Even if some rehabilitation occurred, most ex-convicts are unable to obtain stable employment upon release. They then return to crime to survive.

Mental health and substance abuse - Many criminals have mental health and substance abuse issues that are not treated appropriately in a prison environment. Those remain when they are released.

Lack of support for new behaviors - Released prisoners need to be surrounded by positive people who support their efforts to reform. This often requires a new environment and economic support.

Three Strikes: Hidden Chains

In an attempt to provide a quick solution to increasing repeat offenders in the 1990s, some U.S. states adopted the "Three Strikes" laws. These laws mandated life sentences for anyone convicted of three serious crimes, even when the third crime was non-violent[10]. The goal was to keep repeat offenders off the streets permanently.

Leandro Andrade had a history of two previous serious, non-violent crimes and had been imprisoned both times. He was a father and army veteran. In 1995, he was arrested in California for shoplifting nine children's videotapes that were worth $153.54.

Because of his prior convictions, and the new Three Strikes Law, he was sentenced to two consecutive terms of 25 years to life in prison[11].

This is a good example of a solution that resulted in unintended consequences, which included:

1. Because different states' Three Strikes laws differed in severity, with California having the most severe penalties, criminals migrated to states with less severe penalties. States were unprepared for this criminal influx.

2. The cost of incarcerating individuals for life, especially for non-violent crimes, placed significant burden on state budgets. A legislative analysis of Three Strikes showed the country's operating costs of this program were close to one-half billion dollars per year.

3. The Three Strikes program, while resulting in reducing overall crime levels, also resulted in higher levels of violent crimes. This occurred because the penalty difference for crime severity was flattened[12].

Halden Prison's Quiet Revolution

Interior in Halden Prison by Justis- og beredskapsdepartementet (JD)[13]

Halden Fengsel (Norwegian for Halden Prison) is a maximum-security prison in Halden, Norway. It is the third largest Norwegian prison and was established in 2010. The prison is designed to simulate life on the outside and focuses on preparing the prisoners to reintegrate into society on release[14].

Norway's approach to incarceration emphasizes rehabilitation over punishment and stands as a model for prison reform. Halden Prison's philosophy about its prisoners is that they have done bad things, but they are not bad people.

The prison has no concrete exercise yard. It is like a college campus—surrounded by open grounds and birch and pine trees. A large concrete wall surrounds the entire complex, but there are no guns, razor-wire, or towers.

Prisoners have private rooms with normal furniture, showers, refrigerators, and flat-screen TVs. The windows are unbarred.

They are locked in their rooms 12 hours per day compared to a U.S. prisoner who is typically locked in a cell for 23 hours per day.

The prison provides meals, however, prisoners can purchase ingredients and cook their own meals. Each prisoner is given 53 kroner per day if they leave their cell and either work or participate in activities.

You might think this is a low-security prison. It is not. It is one of Norway's highest security prisons, housing pedophiles, murderers, and rapists.

The goal or purpose of Halden Prison is to ensure that inmates can reintegrate into society effectively. Respectful treatment and guards who serve more as social workers are key elements in this system. Inmates receive education and vocational training in an environment meant to mimic the outside world[15].

So far, Halden's results are promising. The country's recidivism rate is low, with only 20% of released prisoners reoffending within two years compared to the U.S. rate of 76% reoffending within five years[16].

Halden's Prison is a model unlike anything we have seen in the U.S., where rehabilitation efforts exist, but within a closed, barred, and angry environment, with many clashes between inmates and guards and little access to sunlight and free movement.

Understanding Halden Prison With Systems Thinking

To understand Halden Prison's success, we have to look at it as a component of a larger whole of societal values, policies, and outcomes. Systems thinking highlights the interconnectedness of parts within a whole and emphasizes how changes in one part of a system can influence other parts. Let's look at Halden Prison through the systems thinking lens.

Halden Prison's design and function emulate a holistic grasp of incarceration. A holistic philosophy of human rights and dignity imbues the rehabilitation program. The goal of this prison system is to reintegrate prisoners successfully into society. Successful reintegration benefits both the prisoner and the community. The overarching philosophy of human rights and dignity influences the system's policies, inmate treatment, inmate behavior, and guard training.

Halden is a resilient system, continuously adapting to modern knowledge of psychology and criminology. Feedback loops support resiliency by allowing the system to self-regulate and evolve. The low recidivism rate of Norwegian prisoners is a positive reinforcing feedback loop, showing the system is effective. As more released prisoners successfully reintegrate without reoffending, it reinforces the value of the human dignity-centered rehabilitation approach.

A leverage point of respectful behavior toward the prisoners seems to have borne significant results in the Halden system. An emergent property of this prison system is an atmosphere of trust, respect, and focus on rehabilitation.

We speak about system boundaries. Could this prison model work in the United States or other countries? Differences in culture, policy, and politics might not yield the same results.

Halden Prison exemplifies how designing a system for the interconnectedness of all parts leads to a positive outcome. It highlights the importance of taking a holistic view, identifying leverage points, and understanding feedback loops to create humane and effective solutions.

As successful as the Halden Prison system appears, there can be unintended consequences as in any system. We will examine a few.

Excessive cost - Halden Prison spends $90,000 annually to house each prisoner[17]. This is at least three times what the U.S. spends per inmate. Norwegians believe it is worth the reduced recidivism rate, but the system may not be sustainable in tight financial times.

Overemphasis on the in-prison environment - It appears that the focus on the living environment while in prison could be underestimating the importance of post-release support.

Complacency - Prison and governmental officials may believe the current system works well and fail to look for continuous improvement. They might also fail to monitor for signs of weakness that need to be addressed.

Public perception - Some members of the public perceive that prisoners are being given too much comfort and respect and not enough punishment.

Overall, the Halden Prison system has successfully reduced recidivism and has positively integrated released prisoners into society, but it is important to always remember to consider unforeseen consequences.

Evolving Our Inner Circle

Barcelona Water by Biel Morro

In the vast domain of societal dynamics, it is easy to be overwhelmed by the magnitude of change that is needed. Real transformation, however, often begins at the smallest levels—areas within our personal orb of influence. We all have an inner circle comprising family and friends. This inner circle reflects values and a small world where we can influence new paradigms.

An example is Bea Johnson, who wrote a blog called Zero Waste Home and lived a lifestyle that produced almost zero trash. Through her blog and book, she reached beyond her small community and inspired others to think about their consumption and waste habits. Over time, the zero-waste movement saw astonishing growth. What began as a family effort expanded to local community, country, and then globally[18].

By evolving our small inner circle, we create a ripple effect that influences broader social change. In this section, we will explore how to influence our immediate social groups and how to apply change management concepts in this process. We begin with the four-stage change process.

Four-Step Blueprint To Transformation

According to Stroh[19], people are motivated to change when they can articulate what they want and see the discrepancy between their desire and current situation.

At a group level, a common goal is required with a shared set of values. Not only do they need to see the difference between where they are and where they want to be, but they also need a shared understanding of why that difference exists. To understand the "why," the group must share a holistic understanding of the problem.

Stroh puts forward four steps to take in starting and motivating a collective group for a shared cause.

1. Establish a change foundation and secure group commitment.

2. Assess the present situation, noting events, trends, and root structures.

a. Diagram the system.

b. Zoom out to see the system within a larger system.

c. Analyze the system for bottlenecks and leverage points.

3. Prioritize the desired goal, making an explicit choice to pursue the solution.

a. Explore solutions.

b. Add stakeholders for new experiences.

4. Target key interventions that offer substantial progress with minimal shifts.

a. Rewire cause-effect relationships.

b. Implement continuous learning and outreach.

Let's look at an example of a small group's work that led to significant global change.

Wangari Maathai: Kenya's Green Beacon Of Hope

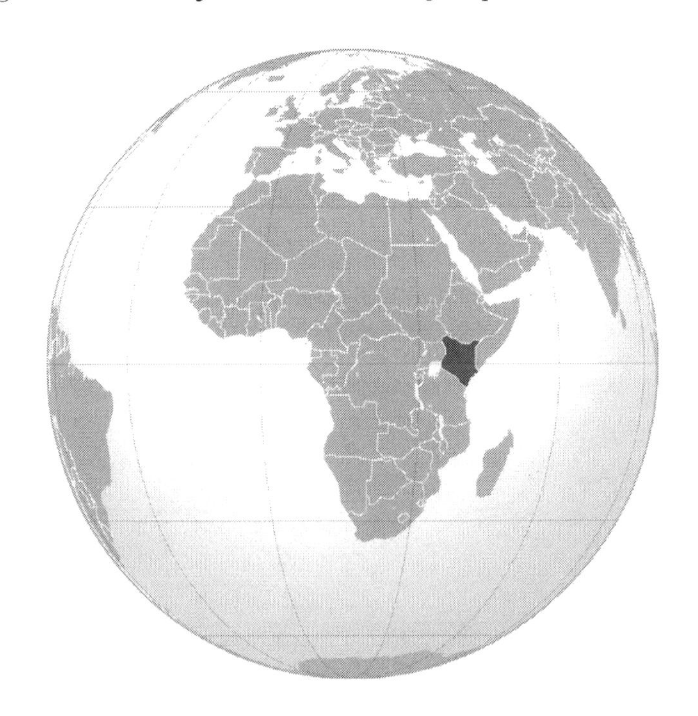

Orthographic Projection Map of Kenya Highlighted in Green by Mandingoesque[20]

In the heart of the 1970s, Kenya faced an environmental crisis. Deforestation had stripped the land, leading to dwindling water sources, vanishing firewood, and eroding soil. For rural communities, the loss of firewood wasn't just about trees; it was a blow to their livelihood, their primary energy source.

Enter Wangari Maathai, a visionary Kenyan environmentalist. She saw beyond the barren landscapes, recognizing the intricate web connecting deforestation to societal challenges. Gathering a passionate group of local women, Wangari ignited a simple yet profound mission: plant trees and rejuvenate the essence of their communities. This wasn't just about greenery; it was a movement to reclaim resources, empower voices, and revitalize community spirit.

But the roots of the problem ran deep. Beyond the visible scars of deforestation lay intertwined issues of governance, land rights, women's rights, and education. A mere replanting wouldn't suffice; a holistic renaissance was needed.

From these humble beginnings, the Green Belt Movement (GBM) blossomed. Spreading its branches across Kenya, the GBM empowered women to cultivate tree nurseries, offering them not just income but a renewed sense of purpose. Their efforts were twofold: restoring the environment and weaving a tapestry of education that championed ethical governance, community unity, and the pivotal role of women in conservation.

Today, with over 50 million trees standing tall in Kenya, the GBM stands as a testament to grassroots conservation and community empowerment. And at its heart, Wangari Maathai, whose dedication earned her the Nobel Peace Prize, remains a beacon of hope and transformation[21].

Similar to the Halden Prison example, the Green Belt Movement is largely successful in its goals of conservation, empowerment, and community sustainability. However, there have been the following unintended consequences:

Political backlash - In the area of land rights and conservation, there were conflicts with political and economic forces in Kenya. Wangari and her group were harassed and faced violent encounters.

Species selection - Though the movement prioritized indigenous trees, non-native species were sometimes selected, which might not have been optimal for the ecosystem. However, the variety of species contributed positively to the farmers' diets and to increasing biodiversity[22].

The unintended consequences typically result from not appreciating the intricacy of the involved infrastructures, institutions, and perceptions[23].

Despite the challenges faced, through Wangari Maathai's story, we see that impactful changes can start with your small inner circle. Her story reveals key elements in successfully harnessing your inner circle for a cause:

1. Perceive the bigger picture, seeing interconnections between and within systems.

2. Don't focus on quick fixes—look at the situation/system in depth.

3. Create long-term solutions first, then short-term solutions that support the long-term. Consider the potential unintended consequences.

Conclusion

In the delicate tapestry of life, every thread holds significance, and every connection matters. From the vast forests of Kenya to the intimate gatherings around family dinner tables, the lessons of interconnectedness echo with clarity.

Aria's discovery of the spider web, the shimmering emblem of nature's intricate design, serves as a poignant reminder of our own societal webs. Each action, each decision, and each whisper of change reverberates through the vast expanse of these webs, touching lives in ways both seen and unseen.

The stories shared in this chapter, from Wangari Maathai's Green Belt Movement to the transformative approach of Halden Prison, illuminate the power of systems thinking. They showcase the potential of small groups, driven by passion and purpose, to instigate monumental shifts. But they also caution us about the complexities and unintended consequences that arise when we tug at the threads of these systems.

As we close this chapter, let's carry forward the wisdom of Mr. Elliot and young Aria. Let's be ever mindful of our actions, understanding that even the faintest whispers can create ripples of change. For in the dance of systems, every step, every gesture, every whisper matters. The challenge now is not just to understand but to act, to weave our own threads into the vast web of society, and to do so with intention, compassion, and foresight.

Action Steps

Is there a cause that you are passionate about? Maybe an issue in your city is bothering you, and you would like to work toward a change. How would you identify a small group of like-minded people to work with you?

Be prepared to thoroughly research the issue you are examining. With research in hand, show your group the current situation, then brainstorm with the group to understand where you want to be—what the goal is. In this small group, create a stock and flow diagram depicting the relevant system. Next, zoom out from that system to see the surrounding systems that might influence behaviors in the system you want to change.

Explore solutions, looking for leverage points and ways to rewire cause and effect loops. Empower the group to provide education and outreach to others in the community.

Chapter Summary

1) Societal and social systems are complex and interwoven, with visible and invisible parts.

2) Quick-fix solutions will not work for complex social issues.

3) Identify long-term solutions first with short-term solutions that support the long-term goals.

4) It is important to identify the current reality of a situation and to understand the discrepancy between that reality and the desired end result.

5) Remember to use a wide-angle lens to see the system within the larger whole.

6) When solutions are introduced into a system, there will always be some unintended consequences, and it is important to identify those, working toward continuous improvement and identifying new leverage points.

7) Your inner circle can initiate small changes in family, work, or community issues using a four-step change process.

8) The four-step change process includes:

a. Securing a small group's commitment to improve an issue.

b. Assessing the present situation through systems thinking.

c. Ensure that the group makes an explicit choice towards the solution.

d. Target key interventions that offer substantial progress with minimal shifts.

8

HOW TO DANCE WITH SYSTEMS

YOUR GUIDE TO INTEGRATING SYSTEMS THINKING PRINCIPLES AND HONING YOUR HOLISTIC DYNAMIC THINKING SKILLS

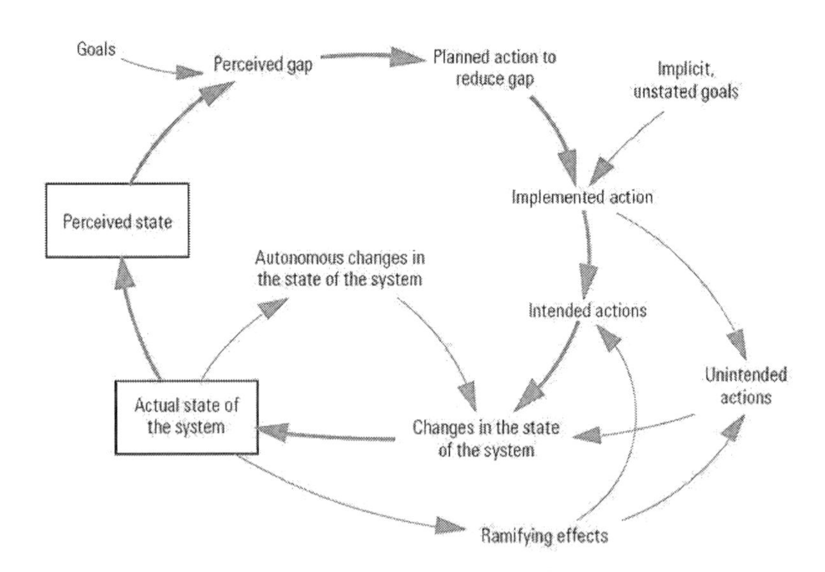

Feedback Loops in a System Dynamics Model by National Cancer Institute

In a suburban neighborhood, residents were increasingly concerned about cars speeding through their streets. Children played in the area, and the neighborhood had no sidewalks. A significant safety issue existed. Neighbors discussed the issue and decided to request the city to install speed bumps or more stop signs.

One neighbor, Adam, who had a background in systems thinking and urban planning, asked to take a broader view of the issue. Instead of focusing only on speeding cars, he wanted to understand the system leading to that behavior.

Adam began by interviewing drivers and observing traffic patterns. He talked to those living in the neighborhood and those just passing through. He took the time to map out the wider road map and included parks, schools, shopping center, and other points of interest.

Through this research, Adam saw several interconnected problems:

1. A close main road became congested during rush hour, which was why drivers used the neighborhood as a shortcut.

2. There were no clear pedestrian crosswalks, making it unclear where children and others might cross the road.

3. The neighborhood roads were wide and straight, encouraging higher speeds.

Adam brought the neighbors together and proposed the following holistic solution:

1. Collaborate with the city to improve traffic flow on the main road, reducing driver's desire to take the shortcut.

2. Paint visible crosswalks in the neighborhood.

3. Implement traffic-calming measures in the neighborhood, like narrowing the roads with the use of planters.

4. Create natural play areas in the neighborhood to alert drivers that children might be present.

5. Launch a community awareness campaign about safe neighborhood driving emphasizing children's safety and the importance of driving slowly.

The neighbors formed a group to work with the city and to implement Adam's ideas for the neighborhood. The speeding issue was significantly reduced. As an added benefit, the neighborhood was more aesthetic, as well as pedestrian and child-friendly.

1. Can you think of other solutions that might have reduced speeding in this neighborhood?

2. Could there be unintended consequences to the neighborhood group's solutions?

Every day, we navigate a myriad of decisions, from the seemingly mundane—like choosing a route to work—to the more complex, such as managing work-life balance or resolving conflicts. Behind each of these decisions lies a web of interconnected factors, a system that often goes unnoticed.

But what if, like Adam, we could see the patterns? What if, in our daily routines and personal challenges, we could apply a lens that helps us understand and optimize these systems for better outcomes?

This chapter is your guide to doing just that. Here, we'll explore how systems thinking, often associated with large-scale global issues, is equally relevant and transformative in our everyday lives. Through relatable examples and actionable insights, this chapter will empower you to apply systems thinking in your personal and professional spheres, turning everyday challenges into holistic successes.

The Dance Of Dynamic Thought

Meadows, described a set of "system wisdoms[1]," principles and practices that guide systems thinking, which are based on the non-linearity, self-organizing, and feedback-driven nature of systems. The "wisdoms" are the dance moves in systems thinking:

1. Capture the system's rhythms.
2. Challenge and clarify your mental models.
3. Value, respect, and disseminate information.
4. Communicate with specific, concise, and truthful language.
5. Balance numeric data with qualitative insight.
6. Implement feedback with clear policies.
7. Optimize whole system properties, not just system parts.
8. Trust the system's innate ability to organize itself.
9. Ensure a system has intrinsic accountability.
10. Commit to lifelong learning.
11. Revere complexity.
12. Expand time vistas
13. Avoid the trap of academic disciplines.

14. Embrace global responsibility.

15. Avoid settling for subpar performance.

16. Let's look at this guidance in detail and with examples.

Syncing With The System's Beat

To understand a system and eventually intervene to improve it, you must first grasp its behavior. What has the system's behavior been over time? Interview those who interact with the system, and look at the system's numbers and other outputs. Create a timeline or graph the data. Ask what the system is doing and why it is doing that.

Capturing The Coffee Shop's Daily Cadence

Sign of Coffee Shop on Ancient Building by Wendy Wei

Imagine that you have been asked to observe the ebb and flow of a coffee shop to understand its rhythm. You are not there as a customer but sitting in a corner watching the motions of the coffee shop system. Some patterns emerge:

Morning rush (6 a.m. to 9 a.m.) - A surge of customers, primarily downtown workers, get their morning coffee. The baristas are working fast and continuously. There is a steady flow of customers in and out and a steady outflow of coffee and pastries at the takeout counter. Seats are empty.

Mid-morning (9:30 a.m. to 11:00 a.m.) - The takeout line slows, and the staff restocks supplies, cleans counters and appliances, and preps for the lunch crowd. Several tables are occupied by remote workers and students with their laptops.

Lunch (12:00 p.m. to 1:30 p.m.) - This is a smaller surge than the morning rush and customers purchase more food items and fewer coffee items. The seating area is full, occupied by groups and single individuals.

Afternoon (2:00 p.m. to 4:30 p.m.) - Very slow traffic in and out, a few customers at the tables. Staff again restocks and cleans. Customers and staff are relaxed and talkative.

Evening buzz (5:00 p.m. to 7:00 p.m.) - Customer flow is steady, with more groups coming in. There is a greater demand for herbal teas and light snacks. Customers are a mix of takeout and eating/drinking.

Knowing the rhythm of the coffee shop leads to effective decisions in staffing, product purchase selections, inventory management, and effective promotion strategies.

Let's stay with our example of you as an observer in the coffee shop, attempting to understand the rhythm of this business system. Remember our discussion of personal mental models? You might have several pre-existing mental models that could influence your perceptions of the coffee shop[2]. For example, you might believe, based on your own preferences and your culture, that espresso-based drinks will be the most popular. However, your observation shows otherwise. Many customers prefer drip coffee, and in fact, it is the most popular order every day that you have observed.

How can you bring these hidden mental models to light?

As part of your role as observer, you can create two columns on a piece of paper. In the first, write your expectations about the customer demographics and customer product preferences (most desired and least ordered). In the second column, write what you actually observe throughout the day. This is a great way to reveal your own mental models to yourself and to allow you to challenge your mental model with facts.

True Guardians Of Knowledge

In systems thinking, information is the lifeblood that allows a system to function, adapt, and evolve. Understanding the intrinsic worth of timely, accurate information is essential.

Information must be treated with care, understanding the source and potential biases. Without knowledge sharing, different parts of a system cannot coordinate based on a shared understanding of the current state and future possibilities[3].

Selfless Science: Salk's Gift To Humanity

The importance of knowledge and information dissemination is seen in the story of Joseph Salk and the polio vaccine. In the early 1950s, the world was gripped by the fear of polio, a debilitating disease that primarily affected children, leading to paralysis and sometimes death. The race was on to find a vaccine that could prevent this devastating illness.

Enter Dr. Jonas Salk, a medical researcher and virologist. After years of research, in 1955, Salk and his team announced that they had successfully developed a vaccine against polio. The world rejoiced at this groundbreaking discovery, and Dr. Salk was hailed as a hero.

However, what truly set Dr. Salk apart was his decision regarding the patent for the vaccine. When asked who owned the patent, Salk famously replied that there was no patent, asking if you could patent the sun. By choosing not to patent the vaccine, Salk ensured that it could be produced at a low cost, making it accessible to millions around the world[4].

Dr. Jonas Salk's decision to prioritize the dissemination of knowledge and the well-being of humanity over personal gain exemplifies the essence of being a guardian of knowledge. He valued the information he had, respected its potential to save lives, and ensured that it was shared widely for the greater good.

Crystal Clarity: The Art Of Precise Dialogue

Systems thinking, foundationally, is about understanding the intricate web of interconnections, feedback loops, and dynamics within a system. The inherent complexity of systems requires clear and precise language when describing them. Using concise, truthful language ensures that complexities are communicated with no added confusion[5].

Ambiguous language leads to misinterpretation. With systems thinking, a misunderstanding can result in a downward spiral of effects, leading to misguided interventions.

Truthful communication builds trust. When stakeholders trust the information they receive, they are more likely to engage in systems thinking and proposed interventions.

Systems thinking teams often include experts from diverse fields. Clear language ensures that they communicate effectively, leading to collaborative teamwork. Precise and concise communication fosters improved feedback loops.

The Miscommunication Crisis Of Flint

Flint River by Edward Kobayashi[6]

In 2011, the state of Michigan took over the city of Flint's finances after a 25% deficit was found. Part of the deficit was a water fund shortfall, so the city planned a new pipeline to bring water from Lake Huron to Flint. However, while the pipeline was being built in 2014, the city changed its water source to the Flint River. Quickly, the residents reported changes to the water's color, smell, and taste[7].

The decision surprisingly resulted in exposing over 100,000 of Flint's residents to high lead levels. It was later discovered that there had been no thorough analysis of Flint River's water quality. Precise communication of the risks could have averted this disaster. But the communication crisis continued beyond this.

As residents complained about the water's taste, appearance, and smell, officials did not respond with clear communication. They initially denied and downplayed the issue instead of communicating specifically and truthfully[8].

This lack of truthful communication further angered the residents and eroded their trust. Residents felt that officials had placed cost-savings above their health and safety. Without trust, collaborative solutions could not move forward.

Flint's crisis illustrates the need for truthful, precise communication at every stage, from decision-making to crisis management. Honest and precise communication could have changed the trajectory of the Flint water crisis.

The Dance Of Data And Discernment

In systems thinking, Meadows[9] discusses the importance of not only relying on numeric data but also incorporating qualitative insights. Systems are not just about numbers. They are also about relationships, interconnections, and subtle, nuanced interactions.

To understand a system, stories and experiences matter. While numeric data tells us how many people are buying a new product, only qualitative insight will tell us why they are buying it. Qualitative insight can reveal underlying causes of issues, the motivations of people, or patterns that might not be revealed by numbers alone.

Relying only on numeric data can result in blind spots that then lead to ill-informed decisions. Numbers allow us to understand trends, scale, and magnitude, but qualitative information adds context and richness to our understanding.

Beyond pixels and power: The human touch in tech

Let's look at a hypothetical example of how a tech company might improve a product.

In the early 2000s, a large tech company wanted to improve its smartphone. They had all the quantitative data: screen resolution, battery life, processing speed, and apps that people wanted. Based on these metrics, their phone was superior to the competition. Yet, sales were not reflecting this.

The company conducted customer surveys and focus groups with current and potential customers and with customers using other brands. This qualitative information revealed that, yes, the phone's technical specifications were impressive.

However, in spite of the long battery life, the user interface was seen as clunky and frustrating. Users had trouble finding their apps and navigating their photos. This diminished their overall user experience.

With this added qualitative insight, the company knew where to put its design efforts to make the product more intuitive and usable. The next version of the smartphone achieved a significant sales increase.

The Art Of Structured Response

We have discussed the importance of feedback in keeping a system stable as well as aiding the system in adapting. Feedback is the process where the output or result of a system loops back and influences the input, thus affecting subsequent outputs. Feedback loops can amplify or stabilize a system's behavior[10].

However, feedback loops alone are not enough for a system to function effectively. The system must have clear policies for how the system is to interpret and act upon the feedback received.

Clear policies ensure that misinterpretation does not occur. Feedback policies allow systems to change and adapt in a structured way, ensuring that beneficial adaptation occurs and not just random reactions to immediate stimuli.

Unambiguous policies make it easier to hold entities accountable for actions in response to feedback. With structured policies, the system can learn from successes and failures, leading to continuous improvement[11].

The Feedback Maze: Ecosync's Journey To Clarity

EcoSync Innovations, a rising fictional tech startup, developed a cutting-edge smart home system aimed at minimizing energy wastage. Their system was engineered to adapt to user habits, optimizing heating, lighting, and air conditioning for energy conservation.

After their market debut, EcoSync introduced a feedback feature in their app, inviting users to highlight glitches or propose enhancements. However, they failed to set a distinct policy for processing, ranking, or replying to this feedback.

A few months post-launch, EcoSync was swamped with a deluge of feedback entries. Repeated issue reports from users, coupled with enhancement suggestions already on their roadmap, cluttered their feedback system.

The absence of a structured feedback policy meant urgent issues were occasionally overlooked, resulting in user discontent. Additionally, the silent treatment users received post-feedback submission added to their grievances.

Realizing the ensuing chaos, EcoSync restructured their feedback approach. They segmented feedback into a) Urgent Bugs, b) Feature Recommendations, and c) General Feedback. Each segment was overseen by a specialized team responsible for evaluation and action. Immediate attention was given to urgent bugs, while feature recommendations were assessed based on viability and user demand. Every feedback entry was acknowledged, and users were informed when their ideas were brought to life.

This revamped feedback policy not only enhanced EcoSync's operational efficiency but also fortified their relationship with users. Customers felt their voices mattered, leading to heightened brand commitment and positive endorsements.

Beyond The Parts: The Whole System Symphony

The Dublin Philharmonic Orchestra by Derek Gleeson[12]

Systems thinking's wisdom urges us to look beyond individual system elements and focus on the system as a whole. This principle highlights the idea that the essence and behavior of systems cannot be comprehended or improved by adjusting only their individual parts.

It is like a symphony orchestra. Each instrument has a unique tone and sound, and each plays a role in the overall performance. While each instrument must be tuned individually, only how the instruments are played simultaneously produces the overall symphonic experience.

The Organic Blueprint: Trusting System Autonomy

All systems, whether a business, family, community, or ecosystem, have an inherent ability to organize and reorganize themselves. Rather than imposing external controls and structures on a system, it is often more effective to allow the system to find its own equilibrium. A system self-organizes through the system elements' interactions and their response to feedback.

Potluck Harmony: The Self-Organizing Feast

Eat Alberta Potluck by Mack Male[13]

Potluck dinners have always been a popular form of entertainment! You organize a potluck with friends, family, or community members. Instead of assigning each person to a dish, you ask everyone to bring a dish they would like to share. There is no central control.

Amazingly, on the day of the potluck, a variety of dishes appear, including appetizers, salads, main courses, veggies, desserts, and drinks. One person might remember a friend One attendee, who is a vegetarian and, thinking others might be also, brings a vegetarian dish. Another remembers there was a shortage of bottled water last time and buys extra on the way. With no central planning, the collective group fosters the hope there is a balanced feast for all.

This is a simple but real example of self-organization. Everyone invited to the potluck made individual decisions on what to bring, and the overall potluck system came together balanced and effectively.

The potluck thrives due to the lack of control. The collective effort fostered the diversity of the group.

A Structured Flexibility

Self-organization can occur simultaneously with other systems thinking principles and practices, even if they appear contradictory—for instance, the relationship between self-organization and clear feedback policies.

Self-organization is about the system adapting and adjusting to achieve a desired state. However, without some form of control, it could potentially lead to instability or unintended consequences. The feedback loop, along with predefined policies or rules, serves as a regulatory mechanism. It continuously monitors the system's state, compares it to the desired state, and adjusts accordingly.

Think of it as a checks-and-balances system. Self-organization allows for flexibility and adaptation, but the feedback loop ensures that these adaptations are within certain boundaries defined by the feedback policies. This way, the system can achieve a balance between flexibility and stability, preventing it from spiraling into chaos or reaching undesirable states.

The feedback loop is a crucial component that ensures the self-organizing system remains within desirable parameters. It adds a layer of control to the inherent adaptive nature of self-organization.

Maybe in the potluck example, a signup sheet is added where everyone writes down what they want to bring. Then, by reading the list, they self-adjust, so it is still self-organizing, but the feedback loop also exists.

Embedded Integrity: The System's Inner Compass

It is essential for a system to have inherent accountability built in to establish proper functioning and alignment with its purpose. Intrinsic accountability means the system has mechanisms that naturally guide it toward its desired goals. The system self-regulates and self-corrects to maintain integrity.

In high-functioning work teams, members take responsibility for their actions not only due to external pressure but also because they recognize their value to the higher organizational mission. When this type of internal compass exists in a system, external intervention can be minimized, and the system becomes resilient and able to adapt and remain effective when faced with challenges. In a business, you would see a culture where taking correct action is the natural course of action.

The Unified Force Of Shared Responsibility

At an exclusive consulting firm, the owners decided to do away with the traditional hierarchy of junior, senior, and lead consultants. Everyone was called a consultant.

Every project was assigned to a team but there was no designated leader. Each member was encouraged to understand all aspects of the client's needs. Weekly client deep-dives were implemented, and the team collectively examined the client's industry, business, objectives, and challenges. All team members shared the holistic view.

Team members did not stay in their lane in terms of expertise but collaborated across disciplines, jumping in where they felt they could add value. This diversity and fluidity ensured the client received all-inclusive solutions.

Each consultant felt intrinsic responsibility to the client and team, knowing that their individual success was tied to the group's success.

Celebrating System Complexity

Systems should not be oversimplified. It is important that you, as a systems thinker, embrace and respect the intricate and multifaceted nature of systems. Systems are inherently complex with interrelated elements, feedback loops, and dynamic interactions. This allows systems to be sustainable.

Remember that while analytical thinking might zoom in on a system element to understand it, systems thinking zooms out to put a system within an even larger whole to see the vast array of interconnections[14].

Complex systems are unpredictable and result in unexpected emergent behaviors. When we acknowledge the complexity, we prepare for unforeseen outcomes.

Tangled Nets: The Unraveling Of The Northern Cod's Labyrinth

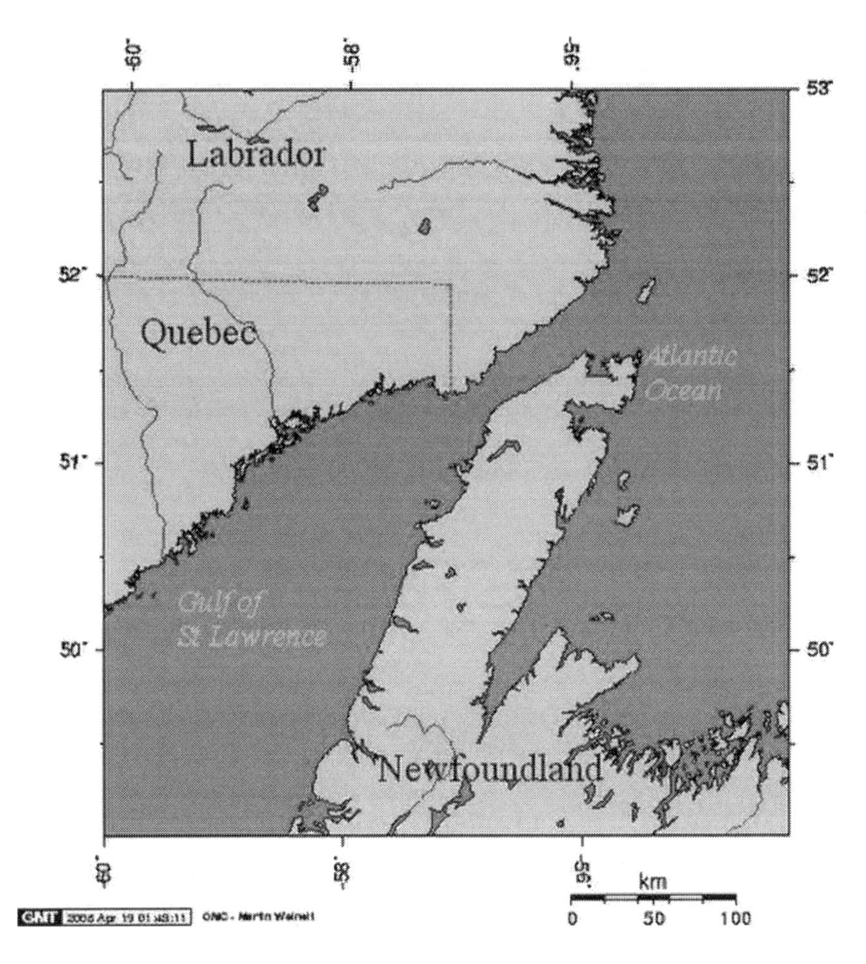

The Strait of Belle Isle by Geo Swan

Flourishing cod populations (called Northern Cod) were observed in the waters around Newfoundland in 1497 by Jon Cabot, an explorer. Starting in the 16th century, this area became a highly productive fishing region. Northern Cod became the reason for European migration to the island. Cod fishing became the primary livelihood of the island population[15].

Northern Cod populations, off the coast of the island of Newfoundland, Canada, dropped to 1% of previous levels in

1992, ending the five-century fishing success story. It had once been thought of as one of the world's richest fisheries. Biological analyses showed overfishing as the primary cause[16]. Environmental conditions were also a contributing factor[17]. Despite a moratorium on Northern Cod fishing in 1992, the stock did not return.

However, as researchers looked deeper into the system, a complexity was revealed that played a significant role in the collapse.

The marine ecosystem had changed in this region. As the cod disappeared, Capelin increased and began to eat the baby cod, exacerbating an already difficult problem. At the same time, water temperatures and salinity were changing, which affected cod reproduction and their food sources.

The moratorium was later lifted and regulations placed quotas on the number of cod that could be caught. However, outdated and incomplete data were relied upon in setting those quotas. The data overestimated the existing cod population. It was determined that even if all fishing ended, recovery would take years, so in 2003, the federal government shut down the Northern Cod fishery.

Additionally, fishing was not only a livelihood for Newfoundland's people but a way of life for the community. There was resistance to accepting the reduction in cod and to acting on the early warnings.

About 37,000 fishermen and fish plant workers lost their jobs during this collapse, requiring them to reeducate for different jobs. Outport populations declined and the population is more dependent on government assistance.

There were unintended consequences that emerged after the close of the fishery. Women's roles changed, birth rates fell, and could not compensate for the out-migration from the island. Young people attend college and are better prepared to move away.

The Island of Newfoundland's population shifted to three sectors: a) those who remained as fishermen, adapting to higher regulation and the new technology-driven fishery, b) those who left for non-fishing careers, and c) the remainder who stayed on the island, supported by government assistance[18].

The cod fishery collapse was due to multifaceted issues in a complex system. Not viewing the system holistically hastened the loss of a once-thriving ecosystem with enormous socioeconomic impacts on the region. Unintended consequences led to a shift in the island's demographics.

Temporal Tapestries: Beyond The Immediate

When applying systems thinking, it is important to understand how the system has evolved and to know that today's decisions will have ripple effects far into the future[19].

Monat & Gannon[20] recommend the use of Behavior Over Time graphs to chart specific system variable values. They suggest this can be a critical first stage in understanding system behavior and element interrelationships.

Behavior Over Time graphs plot the values of pertinent system variables over time. They are often useful first steps in developing an understanding of systemic behavior and of how variables inter-relate. There is an excellent Behavior Over

Time graph that depicts the Northern Cod fish population, shown below.

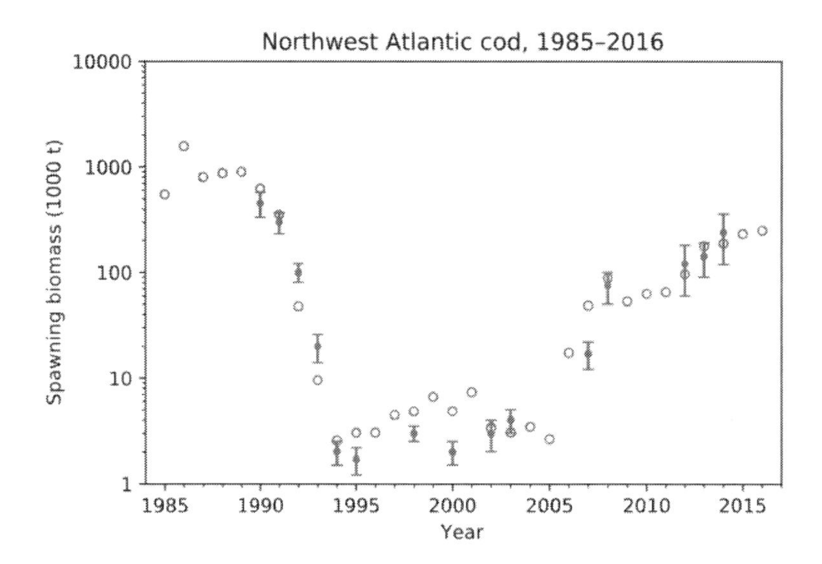

The Collapse and Recovery of the Northwest Atlantic Cod Fishery by Wing gundam[21]

The interconnections in a system are not only spatial but also temporal. Sometimes a decision made in current times will not impact the system for many years. Conversely, a system behavior we see today may have been the result of an intervention from decades ago.

Keeping this temporal perspective in mind, reflect upon how a seemingly simple decision can set in motion a cascade of long-term consequences:

1. Consider when a new road is built in a city. What impacts on the city might occur long-term that can't be seen today? Will the new road lead to new housing in that region?

2. Does the new road foster the use of public transportation or increased use of single cars?

3. Will public transportation use be further ignored because of this new road?

4. Who is benefiting most from this new road?

5. What was the purpose of the road?

6. What were the immediate short-term benefits?

7. Do those short-term goals support the city's long-term goals?

Circles Of Compassion: Beyond Self To System

Systems thinking can profoundly expand your boundary of caring. As you apply systems thinking to challenges, you will move from a narrow focus on self-interest and immediate concerns to consider broader impacts on the entire system[22].

You realize that every small decision made within a system often touches distant parts of the system or parts that may seem unrelated at first glance.

For example, on a personal level, you might begin to think more deeply about the products you purchase or how you interact with others, now knowing you play a role in a much larger whole than you realized. Scenarios that we examined earlier in the book related to recycling and handwashing have far-reaching implications for the health and well-being of all of society.

In a business context, we see companies shifting from focusing on profit only to also caring about social responsibility. Ben & Jerry's, the ice cream company, is a perfect example of this.

Apart from producing delicious ice cream, they actively engage in various social and environmental initiatives.

For instance, Ben & Jerry's sources ingredients like cocoa, vanilla, and sugar through fair-trade practices, ensuring that farmers receive a fair wage. They also support causes like climate justice, racial equality, and GMO labeling. Their dedication to these issues is evident in their flavor names, partnerships, and company campaigns[23].

Pinnacle Pursuit

If we allow subpar performance in any single part of a system, the effect will spread to all interconnected parts, and ultimately the entire system drifts to low performance. The goal in systems thinking is always to optimize every element to ensure the larger system is optimized.

To achieve this, the following is required:

1. Feedback mechanisms that reveal the root causes of underperformance and adapt accordingly.

2. Continuous monitoring of all system elements to detect anomalies and deviations.

3. Creating a culture of continuous learning where mistakes become opportunities to adapt and improve.

4. Establishing unambiguous performance standards and ensuring that all system elements meet or exceed the standards[24].

Lifelong Quest Beyond Silos

To engage in systems thinking, you must commit to a continuous commitment to learning and adapting. In Chapter

7, as we examined a variety of system improvements, we always ended with a discussion of unforeseen consequences that emerged after system changes were implemented. This reflects the continuous learning and adapting required.

In systems thinking, improvements are incremental and meant to be ongoing. Continuous monitoring and feedback continue when the system change occurs, and new challenges will emerge within the system's complexity.

Additionally, systems span multiple disciplines as do systems thinkers. The healthcare system demands knowledge in medicine, social work, economics, and business management, just to name a few. This integrated view protects against issues being overlooked. However, it is always critical to devise feedback loops that allow for constant adjustment to the unforeseen.

Academic disciplines quickly become siloed, viewing issues from a single perspective[25]. It is the responsibility of a systems thinker to bring diverse perspectives to the table.

What If You Ventured Beyond Your Expertise?

Imagine you're a dedicated marine biologist. Your days are filled with studying the delicate ecosystems of coral reefs, and you're recognized in your field for your deep knowledge.

But one day, at a conference, you stumble upon a workshop on systems thinking. It's a transformative experience, making you realize that while you understand marine biology intricately, you might be missing the broader picture.

What if, inspired by this revelation, you decided to take a sabbatical? You travel not just to study marine life, but to immerse yourself in the communities that live alongside these reefs. You dive into courses on economics, social sciences, and technology, expanding your horizons beyond the ocean's depths.

In a coastal town, you collaborate with local fishermen, tech enthusiasts, and economists. Together, you craft a sustainable fishing model, using cutting-edge technology to ensure the reefs and the community's livelihood coexist harmoniously, knowing that challenges remain and continuous monitoring will identify new issues.

Upon your return from sabbatical, your colleagues notice the change. You're not just the marine biologist anymore; you've evolved into a systems thinker, integrating diverse knowledge to address complex challenges.

This journey could be yours. By stepping outside your comfort zone, embracing lifelong learning, and seeing the interconnectedness of all things, you, too, can enhance your approach to systems thinking.

Sharpen Your System's Thinking Savvy

Here are some practical exercises to help you integrate and develop your systems thinking skills.

Using The Language - Flashcards

Objective: Learn the systems thinking language.

Exercise:

1. Create flashcards with systems thinking terms on one side and their definitions on the other.

2. Review these cards daily, testing yourself on both the term and its definition.

3. As you become more familiar, try to use these terms in your daily conversations and observations.

Daily Systems Journal

Notebook Photo By Kelly Sikkema

Objective: Apply the optimal systems thinking ideas, models, and frameworks to day-to-day scenarios. Identify system archetypes.

Exercise:

1. Every day, select a routine scenario or system (grocery checkout, traffic pattern, work process at your job, routine interaction with a friend) - keep it simple.

2. Write about what you observed in your journal, describing the event in terms of systems thinking models.

3. Identify the stocks and flows.

4. Can you recognize a system archetype and related trap?

5. Identify feedback loops, interdependencies, and related system elements.

You will begin to see the world around you through a systems perspective.

Diagramming

Objective: Understand and visualize complex systems.

Exercise:

1. Choose a problem that interests you.

2. Start a mind map with the issue in the center.

3. Branch out with related elements and issues of the central system.

4. Connect the elements with arrows to indicate relationships and feedback loops.

The visual representation allows you to see the bigger picture and understand the system's complexity.

The Art Of Inquisitive Expansion

In your quest to practice systems thinking on a daily basis, ask different questions[26]. Instead of asking about visible events, focus on the hidden relationships, trends, and patterns that are revealed over time. Let's look at a simple example of how to shift the questions you ask.

The Sad Houseplant

Imagine that you have a houseplant in your living room that has wilted and looks unhealthy. Your first instinct is to ask a question that focuses on the event: wilting. You ask, "Did I water the plant today?" Your only focus is on watering. To be safe, you water the plant.

However, the next day, you see the plant is still wilted. Applying systems thinking, you broaden your view and look at the plant system over a longer period[27]. Now, you will ask: How often do I water this plant each week? Have I seen patterns in the plant's health over the last month? How much direct sun does the plant receive every day? Has the room temperature changed?

In taking this wide-angle view of the plant within its surrounding environment, you might discover that the plant is getting too much direct sunlight causing it to dry out quickly. Or, you might realize that you have overwatered the plant for a long time, resulting in root rot.

Declining Productivity

Assume you are a manager in charge of a team whose productivity has declined over the last two months. Your first question might be, "Why didn't the team meet their deadline on the last project?" This question is focused on the event of a missed deadline and is likely to invoke a variety of excuses in response.

Your goal, however, is to improve team productivity, so you need to take a wide view over a longer period and look at the underlying interconnections and less visible structures.

1. How has team morale been over the last quarter?

2. Does productivity dip at particular times of the day?

3. Do team communication patterns align with productivity?

4. What external factors, like changes in other departments or new company initiatives, could be affecting the team?

5. Has a team member left the team?

6. Has a new team member entered[28]?

System Playbooks

Think of the playbooks that football teams use, detailing specific plays and strategies for different game scenarios. In applying systems thinking, you could create a playbook outlining the various models, approaches, and techniques you want to use to address various issues.

Earlier, Ben & Jerry's was used as an example. What does their strategic playbook look like?

<u>Scoop Of Strategy: The Ben & Jerry's Playbook</u>

Ben & Jerry's was founded in 1978 by Jerry Greenfield and Ben Cohen in Vermont. They developed a systematic strategy to integrate community service and environmental responsibility into all aspects of their business.

The Playbook In Action

Environmental Impact - There is a holistic approach to reducing the organization's carbon footprint. Ice cream production, package design, reuse, and recycling are all considerations.

Community Involvement - A portion of Ben & Jerry's yearly profits are provided to community projects. They have implemented a structured process to select these projects.

Ingredient Sourcing - The organization is committed to sourcing non-GMO ingredients, using only fair-trade coffee, bananas, cocoa, sugar, and vanilla. While not an inexpensive method, it ensures they maintain their ethical standards.

Flavors with a Cause - Ben & Jerry's launches some flavors to raise awareness of certain social issues. Justice ReMix'd is an example where racism in the criminal justice system was highlighted[29].

Ben & Jerry's systematic approach toward social responsibility has led them to be profitable and also make a significant positive contribution to communities and the environment. This systems playbook can be a model for other companies entering the realm of social responsibility.

Companies are in a vital position to achieve Sustainable Development Goals (SDGs), but it is difficult for companies to know when they are making effective strategic contributions. SDGs must be removed from their silos so organizations can advance many at the same time[30]. Ben & Jerry's is successful in supporting multiple SDGs simultaneously.

Systematic Playgrounds

Let's have some fun with systems thinking. The following exercises are designed as a playful yet profound gateway into this expansive mindset. By engaging in these activities, you'll not only sharpen your analytical skills but also cultivate an intuitive understanding of systems.

Think of them as mini-adventures for your mind, each one nudging you closer to mastering the art of systems thinking. Dive in, explore, and let your cognitive horizons expand.

The Paper Airplane Factory

Objective: Understand optimization in systems

Instructions:

1. Make a paper airplane.

2. Measure how far it flies.

3. Make a small change to its design (e.g., fold the wings differently).

4. Fly it again and measure.

5. Repeat with various changes.

Reflect: Which changes improved the airplane's flight and which didn't? How does this relate to optimizing systems?

Plant Growth Observation

Objective: Understand the influence of various factors on a system.

Instructions:

1. Plant a seed in a pot.

2. Water it regularly and place it in sunlight.

3. After a week, move it to a dark place and continue watering.

4. Observe the changes for another week.

Reflect: How did changing one factor (light) influence the system (plant growth)? What other factors might influence this system?

Home Energy Audit

Objective: Recognize the energy flow and consumption in your home.

Instructions:

1. Walk around your home and list all the devices and appliances that consume energy.

2. Note how often and for how long each item is typically used in a day.

3. Identify how often each device is used or powers on automatically (e.g., heating systems, refrigerators) and which are used most frequently. You won't know actual consumption, but you will have a high-level view.

4. Reflect on potential ways to reduce energy consumption. Could certain devices be used less often? Are there energy-efficient alternatives?

5. By recognizing patterns and dependencies, you can make informed decisions to optimize energy use and reduce costs.

Reflect: How might this systems thinking approach be applied to other resources in the home, like water or food?

The Supply Chain Game

Objective: Understand the complexities of systems in a business context.

Instructions:

1. Choose a common product, like a pencil or a T-shirt.

2. Research on the Internet and map out its supply chain: Where do the raw materials come from? How is it manufactured? Where is the product stored before distribution? How does it reach the consumer?

3. Also consider how it is distributed from raw material to manufacturer and manufacturer to retail outlet (rail, shipping, air, trucks).

4. Consider potential disruptions at each stage: What if there's a shortage of graphite for the pencil? Or a transportation strike affecting T-shirt delivery?

5. Discuss strategies to mitigate these disruptions and ensure the product reaches the end consumer.

Reflect: How might businesses use systems thinking to better prepare for supply chain challenges?

Conclusion

In our journey through this chapter, we've delved deep into the intricate dance of systems and the principles that guide their harmonious function. We've learned the importance of attuning ourselves to a system's natural rhythms and the value of constantly challenging our preconceived notions.

Information, when respected and disseminated effectively, becomes the lifeblood of a system, and clear, truthful communication is its heartbeat. While numeric data offers a quantifiable perspective, qualitative insights breathe life and context into it.

Feedback, when implemented with clarity, can steer a system towards its optimal path, emphasizing the importance of looking at the system as a cohesive whole rather than disjointed parts.

Trust in a system's organic ability to find order, ensure its inherent accountability, and commit to an ever-evolving journey of learning. In the face of complexity, instead of shying away, we must revere it, broadening our horizons beyond the present and avoiding the confines of narrow academic silos.

Our responsibility is not just local but global, urging us to strive for excellence and never settle for mediocrity. As we move forward, let these principles guide our steps, reminding us of the delicate balance and profound interconnectedness of the world.

Finally, you walk away from the last chapter of this book with fun activities and exercises that you can quickly put to use in your everyday life as you practice systems thinking.

Action Steps

At the end of this chapter, you were given the idea of keeping a systems thinking journal. As an action step, make your first entry into that journal.

1. Pick a simple system you see every day.

2. Choose any way you like to "draw" the system and its interrelated elements.

3. Depict the relationship between elements in any way that works for you.

4. Highlight areas that might represent a delay or bottleneck in the system.

5. Look for leverage points where you might intervene with a simple change that will have large results.

6. Under your drawing, describe the system in words, pointing to challenges in the system that need intervention and your ideas on how that intervention might occur.

7. Share what you created with another person and get their feedback.

Chapter Summary

Systems thinking has "dance moves"—the toolset you need during the systems thinking process. Here is a summary of those moves.

1) Observe how the system functions, seeking the rhythm of the system. Think of the stocks and flows. How are items or events moving in and out of the stocks?

2) Challenge and clarify your personal mental models that might be influencing your interpretations.

3) Value, respect, and disseminate information. Communicate information with specific, concise, and truthful language.

4) Balance numeric data with qualitative insight. Qualitative insight comes from interviews, focus groups, and open-ended surveys.

5) Implement feedback loops with clear policies that clearly state how action is taken in response to feedback.

6) Embrace complexity, expand time vistas, and avoid settling for subpar performance.

7) Optimize whole system properties not just system parts, and don't underestimate the system's innate ability to organize itself.

8) Ensure a system has intrinsic accountability and embraces global responsibility.

9) Commit to lifelong learning, and avoid the trap of academic disciplines.

AFTERWORD

As we draw this exploration to a close, let's circle back to where our journey began: with Isabelle in the heart of a dense forest, captivated by the intricate dance of life unfolding before her. The vibrant ecosystem around the base of a towering pine served as a microcosm of the vast, interconnected world of systems. Just as Isabelle unraveled the secrets of that forest microcosm, we, too, have delved deep into the world of systems thinking, uncovering its principles, intricacies, and applications.

The forest scene painted a vivid picture of nature's systems in action: the symbiotic relationship between moss and decaying wood, the fungi's hidden partnership with the pine tree, the pollinators aiding reproduction, and the ivy's ingenious adaptation. Each element, though distinct, was part of a larger, harmonious whole. This realization mirrors our exploration throughout this book. We've learned that systems are everywhere, from the natural world to our societies, economies, and personal lives. And just like the forest, these systems are interwoven, each part influencing and being influenced by the others.

Isabelle's journey in the forest was a metaphor for our own journey through this book. She left the forest with a newfound appreciation for the systems operating in the natural world, understanding that the designs of her everyday life are equally captivating. Similarly, as we close this chapter, we carry with us a deeper understanding and appreciation for systems thinking. We've equipped ourselves with the tools and insights to navigate the complex web of systems in our world, to see the connections, the feedback loops, and the dynamics at play.

In essence, our exploration has been a symphony of systems thinking, echoing the harmonious dance of growth, balance, and life that Isabelle witnessed in the forest. As we step forth, may we, like Isabelle, continue to see the world with a systems lens, recognizing the interconnectedness of all things and playing our part in shaping a harmonious, sustainable future. Now that you have the systems thinking tools you need, apply them to the large and small challenges that surround you daily. Be a force for change.

OVER 10,000 PEOPLE HAVE ALREADY SUBSCRIBED. DID YOU TAKE YOUR CHANCE YET?

In general, around 50% of the people who start reading do not finish a book. You are the exception, and we are happy you took the time.

To honor this, we invite you to join our exclusive Wisdom University newsletter. You cannot find this subscription link anywhere else on the web but in our books!

Upon signing up, you'll receive two of our most popular bestselling books, highly acclaimed by readers like yourself. We sell copies of these books daily, but you will receive them as a gift. Additionally, you'll gain access to two transformative short sheets and enjoy complimentary access to all our upcoming e-books, completely free of charge!

This offer and our newsletter are free; you can unsubscribe anytime.

Here's everything you get:

✓ How To Train Your Thinking eBook	**($9.99 Value)**
✓ The Art Of Game Theory eBook	**($9.99 Value)**
✓ Break Your Thinking Patterns Sheet	**($4.99 Value)**
✓ Flex Your Wisdom Muscle Sheet	**($4.99 Value)**
✓ All our upcoming eBooks	**($199.80* Value)**

Total Value: $229.76

Go to wisdom-university.net for the offer!

(Or simply scan the code with your camera)

*If you download 20 of our books for free, this would equal a value of 199.80$

THE PEOPLE BEHIND WISDOM UNIVERSITY

Michael Meisner, Founder and CEO

When Michael ventured into publishing books on Amazon, he discovered that his favorite topics—the intricacies of the human mind and behavior—were often tackled in a way that's too complex and unengaging. Thus, he dedicated himself to making his ideal a reality: books that effortlessly inform, entertain, and resonate with readers' everyday experiences, enabling them to enact enduring positive changes in their lives.

Together with like-minded people, this ideal became his passion and profession. Michael is in charge of steering the strategic direction and brand orientation of Wisdom University, as he continues to improve and extend his business.

Claire M. Umali, Publishing Manager

Collaborative work lies at the heart of crafting books, and keeping everyone on the same page is an essential task. Claire oversees all the stages of this collaboration, from researching to outlining and from writing to editing. In her free time, she writes online reviews and likes to bother her cats.

Farley Bermeo, Co-Publishing Manager

Farley has a knack for storytelling and writing personal narratives, both mundane and the extraordinary. Combining his background in writing and experience in program management, he ensures that ideas are transformed into pages. He believes that a good story is better told with a steaming cup of coffee.

Susan Ferebee, Writer

Susan is a professor, researcher, and writer. She holds a Ph.D. in Information Systems. She has published many scholarly articles and book chapters. Susan teaches master's cybersecurity and information systems classes. In her spare time, she enjoys gardening and reading.

Andrew Speno, Content Editor

Andrew is a teacher, writer, and editor. He has published two historical nonfiction books for middle-grade readers, a biography of Eddie Rickenbacker and the story of the 1928 Bunion Derby ultra-marathon. He enjoys cooking, attending live theater, and playing the ancient game of go.

Sandra Agarrat, Language Editor

Sandra Wall Agarrat is an experienced freelance academic editor/proofreader, writer, and researcher. Sandra holds graduate degrees in Public Policy and International Relations. Her portfolio of projects includes books, dissertations, theses, scholarly articles, and grant proposals.

Michelle Olarte, Researcher

Michelle conducts extensive research and constructs thorough outlines that substantiate Wisdom University's book structure. She graduated from Communication Studies with high honors. Her works include screenplays, book editing, book advertisements, and magazine articles.

Alpia Villacorta, Layout Designer

Alpia makes sure that each book follows Wisdom University's formatting and design standards, helping it look outstanding, organized, and reader-friendly. She also helps curate suitable images for book covers. For Alpia, becoming an expert layout designer requires a lot of creativity and attention to detail. She believes that maintaining a positive and joyful attitude, along with reading self-help books, can aid in taking care of one's mental health.

Natalie Briggs, Copywriter

Natalie Briggs is a 20-year veteran of the Caribbean's mediascape, having worked as a journalist, editor, broadcaster, and producer in three countries. In 2020, she turned her hand to copywriting and has worked with Wisdom University for two of those three years. She is a graduate of the University of the West Indies and the University of Leicester. She holds a BA in History and Literature and an MA in PR and Communications.

Jemarie Gumban, Hiring Manager

Jemarie is in charge of thoroughly examining and evaluating the profiles and potential of the many aspiring writers and associates for Wisdom University. With an academic background in Applied Linguistics and a meaningful experience as an industrial worker, she approaches her work with a discerning eye and fresh outlook. Guided by her unique perspective, Jemarie derives fulfillment from turning a writer's desire to create motivational literature into tangible reality.

Evangeline Obiedo, Publishing Assistant

Evangeline diligently supports our books' journey, from the writing stage to connecting with our readers. Her commitment to detail permeates her work, encompassing tasks such as initiating profile evaluations and ensuring seamless delivery of our newsletters. Her love for learning extends into the real world—she loves traveling and experiencing new places and cultures.

REFERENCES

1. What Exactly Is A System?

1. Meadows, D. (2008). *Thinking in systems*. Chelsea Green Publishers.
2. Meadows, D. (2008). *Thinking in systems*. Chelsea Green Publishers.
3. Freepik. (n.d.). *Circulatory system* [Image]. Freepik. https://www.freepik.com/free-vector/gradient-circulatory-system-infographic_10877917.htm
4. Collins, John T. (2019). *Mental models and thinking in systems*. Amazon Digital Services.
5. Collins, John T. (2019). Mental models and thinking in systems. Amazon Digital Services
6. Meadows, D. (2008). *Thinking in systems*. Chelsea Green Publishers.
7. Sliwa, K. (2010). Stock and flow thinking in decision making: Towards systemic procedure of problem solving. *Management Business Innovation, 6*, 52-65.
8. Rutherford A. (2021). *The systems thinker: Essential thinking skills for solving problems, managing chaos, and creating lasting solutions in a complex world. (The systems thinker series, Book 1)*. Independently published
9. Henshaw, M., Dahmann, J., & Lawson, B. (2023, May 17). System of Systems (SoS). SEBoK. In *Media Wiki*. https://sebokwiki.org/wiki/Systems_of_Systems_(SoS)
10. U.S. Department of Defense. (2008). *Systems engineering guide for Systems of Systems*. https://acqnotes.com/wp-content/uploads/2014/09/DoD-Systems-Engineering-Guide-for-Systems-of-Systems-Aug-2008.pdf
11. Meadows, D. (2008). *Thinking in systems*. Chelsea Green Publishers.
12. Rutherford A. (2021). *The systems thinker: Essential thinking skills for solving problems, managing chaos, and creating lasting solutions in a complex world. (The systems thinker series, Book 1)*. Independently published
13. Meadows, D. (2008). *Thinking in systems*. Chelsea Green Publishers.
14. Meadows, D. (2008). *Thinking in systems*. Chelsea Green Publishers.
15. Meadows, D. (2008). *Thinking in systems*. Chelsea Green Publishers.
16. U.S. Capitol. (2023). *About Congress*. Retrieved October 20, 2023, from https://www.visitthecapitol.gov/explore/about-congress
17. Wu, A. (2021, October 19). The Facebook trap. *Harvard Business Review.* https://hbr.org/2021/10/the-facebook-trap
18. Rutherford A. (2021). *The systems thinker: Essential thinking skills for solving problems, managing chaos, and creating lasting solutions in a complex world. (The systems thinker series, Book 1)*. Independently published

19. Rutherford A. (2021). *The systems thinker: Essential thinking skills for solving problems, managing chaos, and creating lasting solutions in a complex world. (The systems thinker series, Book 1)*. Independently published

2. (Extra)Ordinary Encounters With Systems

1. Gleik, J. (2008). *Chaos: Making a new science*. Penguin Books.
2. Environmental Protection Agency. (2023, October 5). *Colony Collapse Disorder*. https://www.epa.gov/pollinator-protection/colony-collapse-disorder
3. Goulson, D., Nicholls, E., Botías, C., Rotheray, E. L., (2015). Bee declines driven by combined stress from parasites, pesticides and lack of flowers. *Science 347* (6229), 1255957. https://doi.org/10.1126/science.1255957
4. Sadler, A. (2016, October 19). Colony collapse disorder: The economics of decline. *California Management Review*. https://cmr.berkeley.edu/2016/10/colony-collapse-disorder/
5. Goulson, D., Nicholls, E., Botías, C., Rotheray, E. L., (2015). Bee declines driven by combined stress from parasites, pesticides and lack of flowers. *Science 347* (6229), 1255957. https://doi.org/10.1126/science.1255957
6. Capra, F., & Luisi, P. (2014). *The systems view of life: A unifying vision*. Cambridge University Press.
7. Capra, F., & Luisi, P. (2014). *The systems view of life: A unifying vision*. Cambridge University Press.
8. Capra, F., & Luisi, P. (2014). *The systems view of life: A unifying vision*. Cambridge University Press.
9. Hardesty, L. (2011, January 19). Prodigy of probability. *MIT News*. https://news.mit.edu/2011/timeline-wiener-0119
10. Capra, F., & Luisi, P. (2014). *The systems view of life: A unifying vision*. Cambridge University Press.
11. Adams, K., Hester, P., & Bradley, J. (2013). *A historical perspective of systems theory*. *IIE Annual Conference and Expo 2013*, 4102–4109. https://docs.edtechhub.org/lib/4P4IZZTP
12. Piryankova, M. (2022). In-depth exploration of systems theory principles: The emergence of self-perfecting systems. *Bulgarian Journal of Business Research, 2*, 13-27.
13. Meadows, D. (2008). *Thinking in systems*. Chelsea Green Publishers.
14. Banzhaf, W. (2009). Self-organizing systems . In: Meyers, R. (Ed.). *Encyclopedia of Complexity and Systems Science*. Springer. https://doi.org/10.1007/978-0-387-30440-3_475
15. Wohlleben, P. (2016). *The hidden life of trees: What they feel, how they communicate - Discoveries from a secret world (The Mysteries of Nature, 1)*. Greystone Books.

16. Simard, S., Perry, D., Jones, M., Myroid, D., Durall, D., & Molina, R. (1997). Net transfer of carbon between ectomycorrhizal tree species in the field. *Nature, 388*, 579-582. https://doi.org/10.1038/41557
17. Toomey, D. (2016, September 1). Exploring how and why trees 'talk' to each other. *Yale Environment 360.* https://e360.yale.edu/features/exploring_how_and_why_trees_talk_to_each_other
18. Simard, S., Perry, D., Jones, M., Myroid, D., Durall, D., & Molina, R. (1997). Net transfer of carbon between ectomycorrhizal tree species in the field. *Nature, 388*, 579-582. https://doi.org/10.1038/41557
19. Charlotte Roy, Salsero35, & Nefronus. (2020). *Mycorrhizal network* [Image]. Wikimedia commons. https://commons.wikimedia.org/wiki/File:Mycorrhizal_network.svg
20. Toomey, D. (2016, September 1). Exploring how and why trees 'talk' to each other. *Yale Environment 360.* https://e360.yale.edu/features/exploring_how_and_why_trees_talk_to_each_other
21. Toomey, D. (2016, September 1). Exploring how and why trees 'talk' to each other. *Yale Environment 360.* https://e360.yale.edu/features/exploring_how_and_why_trees_talk_to_each_other
22. Meadows, D. (2008). *Thinking in systems.* Chelsea Green Publishers.
23. Meadows, D. (2008). *Thinking in systems.* Chelsea Green Publishers.
24. Walker, B., & Salt, D. (2006). *Resilience thinking: Sustaining ecosystems and people in a changing world.* Island Press.
25. Meadows, D. (2008). *Thinking in systems.* Chelsea Green Publishers.
26. Meadows, D. (2008). *Thinking in systems.* Chelsea Green Publishers.
27. Weinberg, G. (2001). *An introduction to general systems thinking.* Dorset House Publishing.
28. Peters, D. (2014). The application of systems thinking in health: Why use systems thinking? *Health Research Policy and Systems, 12*, Article 51. https://doi.org/10.1186/1478-4505-12-51

3. Now, Let's Talk About Systems Thinking

1. Capra, F. & Luisi, P. (2014). *The systems view of life: A unifying vision.* Cambridge University Press.
2. Sinha, R. (2008). Chronic stress, drug use, and vulnerability to addiction. *Ann N Y Acad Sci, 1141*, 105-130. https://doi.org/10.1196/annals.1441.030
3. Meadows, D. (2008). *Thinking in systems.* Chelsea Green Publishers.
4. Meadows, D. (2008). *Thinking in systems.* Chelsea Green Publishers
5. Energsoft. (2019, March 4). Energy storage problems. https://energsoft.com/blog/f/energy-storage-problem

6. Benson, T., Coble, M., & Dilles, J. (2023). Hydrothermal enrichment of lithium in intracaldera illite-bearing claystones. *Science Advances, 9*(35), eadh8183. https://doi.org/10.1126/sciadv.adh8183
7. Birk, A. (2022). Is a battery shortage hampering solar & wind development? *The Environmental Magazine*. https://emagazine.com/is-a-battery-shortage-hampering-solar-wind-development/
8. Shan, R., Reagan, J., Castellanos, S., Kurtz, S., & Kittner, N. (2022). Evaluating emerging long-duration energy storage technologies. *Renewable and Sustainable Energy Reviews, 159*. https://doi.org/10.1016/j.rser.2022.112240
9. Energsoft. (2019, March 4). Energy storage problems. https://energsoft.com/blog/f/energy-storage-problem
10. Shan, R., Reagan, J., Castellanos, S., Kurtz, S., & Kittner, N. (2022). Evaluating emerging long-duration energy storage technologies. *Renewable and Sustainable Energy Reviews, 159*. https://doi.org/10.1016/j.rser.2022.112240
11. Hanley, S. (2023, January 14). Energy storage is going underground. *Clean Technica*. https://cleantechnica.com/2023/01/14/energy-storage-is-going-underground/
12. Shan, R., Reagan, J., Castellanos, S., Kurtz, S., & Kittner, N. (2022). Evaluating emerging long-duration energy storage technologies. *Renewable and Sustainable Energy Reviews, 159*. https://doi.org/10.1016/j.rser.2022.112240
13. Murray, C. (2022). World's first large-scale 'sand battery' goes online in Finland. *Energy Storage News*. https://www.energy-storage.news/worlds-first-large-scale-sand-battery-goes-online-in-finland/
14. Stroh, P. (2015). *Systems thinking for social change*. Chelsea Green Publishing.
15. Shellenberg, M. (2020, August 15). Why California's climate policies are causing electricity blackouts. *Forbes*. https://www.forbes.com/sites/michaelshellenberger/2020/08/15/why-californias-climate-policies-are-causing-electricity-black-outs/?sh=4eb831981591
16. Office of the Federal Chief Sustainability Officer. (n.d.). *Net-zero emissions by 2050, including a 65% reduction by 2030*. Retrieved October 20, 2023, from https://www.sustainability.gov/federalsustainabilityplan/emissions.html
17. Rotsee. (2019). *Samsø, Denmark* [Map image]. Wikimedia Commons. https://en.m.wikipedia.org/wiki/File:Denmark_location_samso.svg
18. Wear, A. (2020, December 25). The island where everyone owns the wind. *Reasons to be Cheerful*. https://reasonstobecheerful.world/the-island-where-everyone-owns-the-wind/
19. Wessman, Johan - NewsØresund. (2015). *Danskt lantbruk och vindkraft på södra Samsø, fotograferat från Vesborg fyr* [Photo]. Wikimedia Commons.

https://commons.wikimedia.org/wiki/File:
Samsoe_Vesborg_20150802_0097_(20680670489).jpg

4. System Secrets

1. Rutherford A. (2021). *The systems thinker: Essential thinking skills for solving problems, managing chaos, and creating lasting solutions in a complex world. (The systems thinker series, Book 1)*. Independently published.
2. B Jana. (2009). *Causal loop diagram - system archetype "Fixes that fail"* [Diagram]. Wikimedia Commons. https://en.wikipedia.org/wiki/File:Fixes_that_fail.PNG
3. Meadows, D. (2008). *Thinking in systems*. Chelsea Green Publishers.
4. Goulson, D., Nicholis, E., Botias, C., & Rotheray, E. (2015). Combined stress from parasites, pesticides and lack of flowers drive bee declines. *Science, 347* (6229). doi: http://dx.doi.org/10.1126/science.1255957
5. B Jana. (2009). *Causal loop diagram - system archetype "Tragedy of the commons"* [Diagram]. Wikimedia Commons. https://commons.wikimedia.org/wiki/File:Tragedy_of_the_commons.PNG
6. Rutherford A. (2021). *The systems thinker: Essential thinking skills for solving problems, managing chaos, and creating lasting solutions in a complex world. (The systems thinker series, Book 1)*. Independently published.
7. Rutherford A. (2021). *The systems thinker: Essential thinking skills for solving problems, managing chaos, and creating lasting solutions in a complex world. (The systems thinker series, Book 1)*. Independently published.
8. Svarnyp. (2011). *Graph to represent the Fixes that fail archetypes behavior* [Graph]. Wikimedia Commons. https://commons.wikimedia.org/wiki/File:Ftfgraph.png
9. Rutherford A. (2021). *The systems thinker: Essential thinking skills for solving problems, managing chaos, and creating lasting solutions in a complex world. (The systems thinker series, Book 1)*. Independently published.
10. Meadows, D. (2008). *Thinking in systems*. Chelsea Green Publishers.
11. Rutherford A. (2021). *The systems thinker: Essential thinking skills for solving problems, managing chaos, and creating lasting solutions in a complex world. (The systems thinker series, Book 1)*. Independently published.
12. Meadows, D. (2008). *Thinking in systems*. Chelsea Green Publishers.
13. Meadows, D. (2008). *Thinking in systems*. Chelsea Green Publishers.
14. Blumenthal, D., Abrams, M., & Nuzum, R.l (2015). The Affordable Care Act at 5 years. *New England Journal of Medicine, 372* (25), 2451–2458. https://www.nejm.org/doi/10.1056/NEJMhpr1503614
15. B Jana. (2009). *Causal loop diagram - system archetype "Success to successful"* [Diagram]. Wikimedia Commons. https://en.m.wikipedia.org/wiki/File:Success_to_the_successful.PNG

16. Meadows, D. (2008). *Thinking in systems*. Chelsea Green Publishers.
17. Arvinth, K. (2015, September 28). VW scandal: Carmaker was warned by Bosch about test-rigging software in 2007. *International BusinessTimes.* https://www.ibtimes.co.uk/vw-scandal-carmaker-was-warned-about-test-rigging-software-2007-1521442
18. Jones, M., Viswanath, O., Peck, J., Kaye, A., Gill, J., & Simopoulos, T. (2018). A brief history of the opioid epidemic and strategies for pain medicine. *Pain Therapy*, 7(1), 13-21. https://doi.org/10.1007%2Fs40122-018-0097-6
19. Meadows, D. (2008). *Thinking in systems*. Chelsea Green Publishers.
20. Meadows, D. (2008). *Thinking in systems*. Chelsea Green Publishers.
21. Rutherford A. (2021). *The systems thinker: Essential thinking skills for solving problems, managing chaos, and creating lasting solutions in a complex world. (The systems thinker series, Book 1)*. Independently published.
22. Meadows, D. (2008). *Thinking in systems*. Chelsea Green Publishers.
23. Rutherford A. (2021). *The systems thinker: Essential thinking skills for solving problems, managing chaos, and creating lasting solutions in a complex world. (The systems thinker series, Book 1)*. Independently published.
24. Meadows, D. (2008). *Thinking in systems*. Chelsea Green Publishers.
25. Meadows, D. (2008). *Thinking in systems*. Chelsea Green Publishers.
26. Rutherford A. (2021). *The systems thinker: Essential thinking skills for solving problems, managing chaos, and creating lasting solutions in a complex world. (The systems thinker series, Book 1)*. Independently published.
27. Meadows, D. (2008). *Thinking in systems*. Chelsea Green Publishers.
28. Los Angeles Times. (1995, September 21). AT&T Breakup II: Highlights in the history of a telecommunications giant. *Los Angeles Times Archives.* https://www.latimes.com/archives/la-xpm-1995-09-21-fi-48462-story.html
29. Meadows, D. (2008). *Thinking in systems*. Chelsea Green Publishers.
30. Metz, R. (2021, October 10). Facebook's success was built on algorithms. Can they also fix it? *CNN Business*. https://www.cnn.com/2021/10/10/tech/facebook-whistleblower-algorithms-fix/index.html
31. Meadows, D. (2008). *Thinking in systems*. Chelsea Green Publishers.
32. Meadows, D. (2008). *Thinking in systems*. Chelsea Green Publishers.
33. World Anti-doping Agency. (n.d.). *The World Anti-Doping Code*. WADA. Retrieved October 18, 2023, from https://www.wada-ama.org/en/what-we-do/world-anti-doping-code
34. Meadows, D. (2008). *Thinking in systems*. Chelsea Green Publishers.
35. Collins, J.T. (2020). *Mental models and thinking in systems*. Fabio Giuliano Stella.
36. Rutherford A. (2021). *The systems thinker: Essential thinking skills for solving problems, managing chaos, and creating lasting solutions in a complex world. (The systems thinker series, Book 1)*. Independently published.

37. Holtrop, J., Scherer, L., Matlock, D., Glasgow, R., & Green, L. (2021). The importance of mental models in implementaton science. *Front Public Health, 9*. 680316. https://doi.org/10.3389/fpubh.2021.680316

5. Navigating The Complexities Of The Big Picture

1. Peterson, C. (2020, July 10). 25 years after returning to Yellowstone, wolves have helped stabilize the ecosystem. *National Geographic.* https://www.nationalgeographic.com/animals/article/yellowstone-wolves-reintroduction-helped-stabilize-ecosystem
2. Meadows, D. (2008). *Thinking in systems*. Chelsea Green Publishers.
3. Rutherford A. (2021). *The systems thinker: Essential thinking skills for solving problems, managing chaos, and creating lasting solutions in a complex world. (The systems thinker series, Book 1)*. Independently published.
4. Meadows, D. (2008). *Thinking in systems*. Chelsea Green Publishers.
5. Lallensack, R. (2019, August 29). The accidental invention of the slinky. *Smithsonian Magazine*. https://www.smithsonianmag.com/innovation/accident al-invention-slinky-180973016/
6. Rutherford A. (2021). *The systems thinker: Essential thinking skills for solving problems, managing chaos, and creating lasting solutions in a complex world. (The systems thinker series, Book 1)*. Independently published.
7. Lallensack, R. (2019, August 29). The accidental invention of the slinky. *Smithsonian Magazine*. https://www.smithsonianmag.com/innovation/accident al-invention-slinky-180973016/
8. Meadows, D. (2008). *Thinking in systems*. Chelsea Green Publishers.
9. Rutherford A. (2021). *The systems thinker: Essential thinking skills for solving problems, managing chaos, and creating lasting solutions in a complex world. (The systems thinker series, Book 1)*. Independently published.
10. Svarnyp. (2011). *Graph to represent the Fixes that fail archetypes behavior* [Graph]. Wikimedia Commons. https://commons.wikimedia.org/wiki/File:Ftfgraph.png
11. Rutherford A. (2021). *The systems thinker: Essential thinking skills for solving problems, managing chaos, and creating lasting solutions in a complex world. (The systems thinker series, Book 1)*. Independently published.
12. www.vensim.com. (2009). *Stock and flow chart model example* [Diagram]. Wikimedia Commons. https://commons.wikimedia.org/wiki/File:Model-example.JPG
13. Hczarn. (2020). *A Generalised Graph of a Predator-Prey Population Density Cycle* [Graph]. Wikimedia Commons. https://en.wikipedia.org/wiki/File:Predator_prey_curve.png

14. Meadows, D. (2008). *Thinking in systems*. Chelsea Green Publishers.
15. Meadows, D. (2008). *Thinking in systems*. Chelsea Green Publishers.
16. Anna Juchnowicz.(2018). *Companion planting* [Drawing]. Wikimedia Commons. https://commons.wikimedia.org/wiki/File:Three_Sisters_companion_planting_technique.jpg
17. U.S. Department of Agriculture (USDA). (2023). *The three sisters of indigenous American agriculture*. USDA. https://www.nal.usda.gov/collections/stories/three-sisters
18. Dobson, A. (2014). Yellowstone wolves and the forces that structure natural systems. *PLOS Biology, 12*(12), Article e1002025. https://doi.org/10.1371/journal.pbio.1002025

6. Beneath The Systems Blueprint

1. Gökhan Tolun. (2017). *Great Barrier Reef Corals* [Photo]. Wikimedia Commons. https://commons.wikimedia.org/wiki/File:Great_Barrier_Reef_Corals.jpg
2. Johnson, J., Maynard, J., Devlin, M., Wilkinson, S., Anthony, K., Yorkston, H., Heron, S., Puotinen, M., & vanHooidonk, R. (2013). *2013 Scientific consensus statement: Chapter 2: Resilience of Great Barrier Reef marine ecosystems and drivers of change.* Reef Water Quality Protection Plan Secretariat. https://www.reefplan.qld.gov.au/__data/assets/pdf_file/0021/46173/scsu-chapter-2-resilience.pdf
3. Unesco. (n.d.). *The Sundarbans*. Unesco World Heritage Convention. Retrieved October 18, 2023, from https://whc.unesco.org/en/list/798/
4. Amruta, P.. (2023, March 20). *Pneumatophore - Environment notes*. Prepp. https://prepp.in/news/e-492-pneumatophore-environment-notes
5. Unesco. (n.d.). *The Sundarbans*. Unesco World Heritage Convention. Retrieved October 18, 2023, from https://whc.unesco.org/en/list/798/
6. Banzhaf, W. (2009). Self-organizing systems . In Meyers, R. (Eds.), *Encyclopedia of complexity and systems science*. Springer. https://doi.org/10.1007/978-0-387-30440-3_475
7. Green, D., Sadedin, S., & Leishman, T. (2008). Self-organization. *Encyclopedia of Ecology (2nd Edition) 1*, 629-636. Elsevier. https://doi.org/10.1016/B978-0-444-63768-0.00696-X
8. Fuchs, C. (2003). The Internet as a self-organizing socio-technological system. *Human Strategies in Complexity Research Paper*. https://dx.doi.org/10.2139/ssrn.458680
9. Semaforo GMS. (2022). *Arpanet Points in the 70s* [Diagram]. Wikimedia Commons. https://en.m.wikipedia.org/wiki/File:Arpanet_in_the_1970s.png

10. Lo, A. (2019). *Adaptive markets: Financial evolution at the speed of thought.* Princeton University Press.
11. Simon, H. (1973). The organization of complex systems. In H.H. Pattee (Ed.), *Hierarchy theory: The challenge of complex systems* (pp. 1-27). George Braziller.
12. Simon, H. (1973). The organization of complex systems. In H.H. Pattee (Ed.), *Hierarchy theory: The challenge of complex systems* (pp. 1-27). George Braziller.
13. CactiStaccingCrane. (2022). *Amazon Rainforest* [Orthographic projection]. Wikimedia Commons. https://en.wikipedia.org/wiki/File:Amazon_rainforest_(orthographic_projection).svg
14. Butler, R. (2020, June 4). *The Amazon Rainforest: The worlds largest rainforest*. Mongabay. https://rainforests.mongabay.com/amazon/
15. Whitmore, T.C. (1998). *An introduction to tropical rain forests.* Oxford University Press.
16. Meadows, D. (2008). *Thinking in systems*. Chelsea Green Publishers.
17. Rutherford A. (2021). *The systems thinker: Essential thinking skills for solving problems, managing chaos, and creating lasting solutions in a complex world. (The systems thinker series, Book 1)*. Independently published.
18. Abson, D., Fischer, J., Leventon, J., Newig, J., Schomerus, T., Vilsmaier, U., von Wehrden, H., Abernethy, P., Ives, C., Jager, N., & Lang, D. (2016). Leverage points for sustainability transformation. *Ambio, 46*, 30-39. https://doi.org/10.1007/s13280-016-0800-y
19. Center for Disease Control and Prevention (CDC). (2022, June 15). *Water, sanitation, and environmentally related hygiene (WASH): Hygiene fast facts*. CDC. https://www.cdc.gov/hygiene/fast-facts.html
20. Global Handwashing Partnership. (n.d.). *Why handwashing: Economic impact.* Global Handwashing Partnership. Retrieved October 18, 2023, from https://globalhandwashing.org/about-handwashing/why-handwashing/economic-impact/
21. Unicef. (n.d.). *Handwashing: The simples way to protect against a range of diseases*. Retrieved October 18, 2023, from https://www.unicef.org/wash/handwashing
22. Mahmood, A., Egan, M., Pervez, S., Alghamdi, H., Tabinda, A., Yasar, A., Brindhadevi, K., & Pugazhendhi, A. (2020). COVID-19 and frequent use of hand sanitizers: Human health and environmental hazards by exposure pathways. *Science of The Total Environment, 742*. https://doi.org/10.1016%2Fj.scitotenv.2020.140561
23. Nakis, J. (n.d.). *Does hand washing have a downside?* Health enews. Retrieved October 18, 2023. https://www.ahchealthenews.com/2015/03/12/hand-washing-has-a-downside/
24. DTMMix. (2017, January 20). *A complete guide to recycling history*. https://dtmmix.co.uk/blog/recycling-history/

25. ASM Metal Recycling. (2022, January 27). The economic effects of reduced aluminium can use. https://www.asm-recycling.co.uk/blog/the-economic-effects-of-reduced-aluminium-can-use/
26. United States Environmental Protection Agency. (n.d.). *Summary of the Resource Conservation and Recovery Act.* Retrieved October 18, 2023, from https://www.epa.gov/laws-regulations/summary-resource-conservation-and-recovery-act
27. Recycle Nation. (2023). *Government initiatives aim to make recycling easier.* Retrieved October 18, 2023, from https://recyclenation.com/2022/08/government-initiatives-aim-to-make-recycling-easier/
28. DTMMix. (2017, January 20). *A complete guide to recycling history*. https://dtmmix.co.uk/blog/recycling-history/
29. Recycle Nation. (2023). *Government initiatives aim to make recycling easier.* Retrieved October 18, 2023, from https://recyclenation.com/2022/08/government-initiatives-aim-to-make-recycling-easier/
30. Deer, R. (2021, July 8). *Why U.S. cities are ending single-stream recycling.* Roadrunner. https://www.roadrunnerwm.com/blog/why-cities-are-ending-single-stream-recycling
31. ASM Metal Recycling. (2022, January 27). The economic effects of reduced aluminium can use. https://www.asm-recycling.co.uk/blog/the-economic-effects-of-reduced-aluminium-can-use/
32. Rinkesh. (n.d.). *The three R's: "Reduce, reuse, recycle": Waste hierarchy to enjoy trash free life.* Conserve Energy Future. Retrieved October 18, 2023, from https://www.conserve-energy-future.com/reduce-reuse-recycle.php

7. Break The Mold

1. Stroh, D. P. (2015). *Systems thinking for social change.* Chelsea Green Publishing.
2. Whiteman, G., Walker, B., & Perego, P. (2013). Planetary boundaries: Ecological foundations for corporate sustainability. *Journal of Management Studies, 50*(2), 307-336. https://doi.org/10.1111/j.1467-6486.2012.01073.x
3. Stroh, D. P. (2015). *Systems thinking for social change.* Chelsea Green Publishing.
4. Sawyer, W., & Wagner, P. (2023). *Mass incarceration: The whole pie 2023* [Press release]. Prison Policy Initiative. https://www.prisonpolicy.org/reports/pie2023.html
5. Sawyer, W., & Wagner, P. (2023). *Mass incarceration: The whole pie 2023* [Press release]. Prison Policy Initiative. https://www.prisonpolicy.org/reports/pie2023.html

6. Tim Rodenberg. (2012). *Rikers Island Jail* [Photo]. Flickr. https://www.flickr.com/photos/sheriffaj/8072743603/in/photolist-dimVqx
7. Gonnerman, J. (2014, September 29). *Before the law.* The New Yorker. https://www.newyorker.com/magazine/2014/10/06/before-the-law
8. Durose, M., & Antenangeli, L. (2021). *Recidivism of prisoners released in 34 states in 2012: A 5-year follow-up period (2012-2017).* U.S. Bureau of Justice Statistics. https://bjs.ojp.gov/library/publications/recidivism-prisoners-released-34-states-2012-5-year-follow-period-2012-2017
9. Durose, M., & Antenangeli, L. (2021). *Recidivism of prisoners released in 34 states in 2012: A 5-year follow-up period (2012-2017).* U.S. Bureau of Justice Statistics. https://bjs.ojp.gov/library/publications/recidivism-prisoners-released-34-states-2012-5-year-follow-period-2012-2017
10. Stanford Law School. (n.d.). *Three strikes basics.* Retrieved October 18, 2023, from https://law.stanford.edu/three-strikes-project/three-strikes-basics/
11. Canon, D. (2021, November 1). This army vet and father of three got two life sentences for stealing movies for his kids. *I Taught the Law.* https://medium.com/i-taught-the-law/this-army-vet-and-father-of-three-got-two-life-sentences-for-stealing-movies-for-his-kids-d814ef00cf4f
12. Iyengar, R. (2008). I'd rather be hanged for a sheep than a lamb: The unintended consequences of 'Three-Strikes' laws. *National Bureau of Economic Research,* Working Paper 13784. http://www.nber.org/papers/w13784
13. Justis- og beredskapsdepartementet (JD). (2010). *Interior in Halden Prison* [Photo]. Wikimedia Commons. https://en.m.wikipedia.org/wiki/File:Interior_in_Halden_prison.jpg
14. Gentleman, A. (2012, May 18). Inside Halden, the most humane prison in the world. *The Guardian.* https://www.theguardian.com/society/2012/may/18/halden-most-humane-prison-in-world
15. Kofman, J. (2015, May 31). In Norway, a prison built on second chances. *NPR.* https://www.npr.org/sections/parallels/2015/05/31/410532066/in-norway-a-prison-built-on-second-chances
16. Janzer, C. (2019, February 2022). North Dakota reforms its prisons, Norwegian style. *U.S. News & World Report.* https://www.usnews.com/news/best-states/articles/2019-02-22/inspired-by-norways-approach-north-dakota-reforms-its-prisons#:~:text=The%20Norwegian%20prison%20system%20boasts,per%20100%2C000%20in%20the%20U.S.
17. Kofman, J. (2015, May 31). In Norway, a prison built on second chances. *NPR.* https://www.npr.org/sections/parallels/2015/05/31/410532066/in-norway-a-prison-built-on-second-chances
18. Bate, M. (2018, July 25). Bea Johnson: The zero waste lifestyle. *Matters.* https://mattersjournal.com/stories/zerowaste

19. Stroh, D. P. (2015). *Systems thinking for social change*. Chelsea Green Publishing.
20. Mandingoesque. (2010). *Orthographic projection map of Kenya highlighted in green* [Orthographic projection]. Wikimedia Commons. https://commons.wikimedia.org/wiki/File:Kenya_(orthographic_projection).svg
21. Unesco. (n.d.). *Wangari Maathai: Biography*. Retrieved October 18, 2023, from https://en.unesco.org/womeninafrica/wangari-maathai/biography
22. Gregory, R. (2017, June). *Kenya - The Green Belt Movemen*t. The EcoTipping Points Project. https://ecotippingpoints.org/our-stories/indepth/kenya-tree-planting.html
23. Kronenberg, J., Andersson, E., Barton, D. & Borgstrom, S. (2021). The thorny path toward greening: Unintended consequences, trade-offs, and constraints in green and blue infrastructure planning, implementation, and management. *Ecology and Society, 26*(2), 36. http://dx.doi.org/10.5751/ES-12445-260236

8. How To Dance With Systems

1. Meadows, D. (2008). *Thinking in systems*. Chelsea Green Publishers.
2. Meadows, D. (2008). *Thinking in systems*. Chelsea Green Publishers.
3. Meadows, D. (2008). *Thinking in systems*. Chelsea Green Publishers.
4. Salk Institute for Biological Studies. (n.d.). *History of Salk*. salk. Retrieved October 18, 2023, from https://www.salk.edu/about/history-of-salk/jonasalk/
5. Meadows, D. (2008). *Thinking in systems*. Chelsea Green Publishers.
6. Edward Kobayashi. (2009). *Flint River* [Photo]. Flickr.
7. CNN Editorial Research. (2022, December 13). Flint water crisis fast facts. *CNN.* https://www.cnn.com/2016/03/04/us/flint-water-crisis-fast-facts/index.html
8. Denchak, M. (2018, November 8). Flint water crisis: Everything you need to know. *NRDC.* https://www.nrdc.org/stories/flint-water-crisis-everything-you-need-know
9. Meadows, D. (2008). *Thinking in systems*. Chelsea Green Publishers.
10. Rutherford A. (2021). *The systems thinker: Essential thinking skills for solving problems, managing chaos, and creating lasting solutions in a complex world. (The systems thinker series, Book 1)*. Independently published.
11. Meadows, D. (2008). *Thinking in systems*. Chelsea Green Publishers.
12. Derek Gleeson. *The Dublin Philharmonic Orchestra* [Photo]. Wikimedia Commons. https://commons.wikimedia.org/wiki/File:Dublin_Philharmonic_Orchestra_performing_Tchaikovsky%27s_Symphony_No_4_in_Charlotte,_North_Carolina.jpg

13. Mack Male. (2012). *Eat Alberta potluck* [Photo]. Flickr. https://flickr.com/photos/24311648@N00/7072671637
14. Meadows, D. (2008). *Thinking in systems*. Chelsea Green Publishers.
15. Hamilton, L.C., Haedrich, R.L., & Duncan, C.M. (2004). Above and below the water: Social/ecological transformation in northwest Newfoundland. *Population and Environment, 25* (3): 195–215. https://doi.org/10.1023/B:POEN.0000032322.21030.c1
16. Hutchings, J. A., & Myers, R. A. (1994). What can be learned from the collapse of a renewable resource? Atlantic cod, Gadus morhua, of Newfoundland and Labrador. *Canadian Journalof Fisheries Aquatic Science, 51*, 2126–2146. https://doi.org/10.1139/f94-214
17. Drinkwater, K. F. (2002). A review of the role of climate variability in the decline of northern cod. *American Fisheries Society Symposium, 32*, 113–130.
18. Hamilton, L.C., Haedrich, R.L., & Duncan, C.M. (2004). Above and below the water: Social/ecological transformation in northwest Newfoundland. *Population and Environment, 25* (3): 195–215. https://doi.org/10.1023/B:POEN.0000032322.21030.c1
19. Meadows, D. (2008). *Thinking in systems*. Chelsea Green Publishers.
20. Monat, J. & Gannon T. (2015). What is systems thinking? A review of selected literature plus recommendations. *International Journal of Systems Science, 4*(1), 11-26.
21. Wing gundam. (2019). *The collapse and recovery of the northwest Atlantic cod fishery* [Graph]. Wikimedia Commons. https://commons.wikimedia.org/wiki/File:Northwest_Atlantic_cod_biomass.svg
22. Meadows, D. (2008). *Thinking in systems*. Chelsea Green Publishers.
23. Benmelech, E. (2021, August 1). Ben & Jerry's social responsibility: ESG without the G. *Forbes*. https://www.forbes.com/sites/effibenmelech/2021/08/01/ben--jerrys-social-responsibility-esg-without-the-g/?sh=4e3eb7a03084
24. Meadows, D. (2008). *Thinking in systems*. Chelsea Green Publishers.
25. Meadows, D. (2008). *Thinking in systems*. Chelsea Green Publishers.
26. McKey, Z. (2019). *Think in systems.* Independently Published.
27. McKey, Z. (2019). *Think in systems.* Independently Published.
28. McKey, Z. (2019). *Think in systems.* Independently Published.
29. Ben & Jerry's. (n.d.). *Our values, activism and mission*. Retrieved October 18, 2023, from https://www.benjerry.com/values
30. van Zantern, J., & van Tulder, R. (2021). Improving companies' impacts on sustainable development: A nexus approach to the SDGs. *Business Strategy and the Environment, 30*(8), 3703-3720. https://doi.org/10.1002/bse.2835

DISCLAIMER

The information contained in this book and its components, is meant to serve as a comprehensive collection of strategies that the author of this book has done research about. Summaries, strategies, tips and tricks are only recommendations by the author, and reading this book will not guarantee that one's results will exactly mirror the author's results.

The author of this book has made all reasonable efforts to provide current and accurate information for the readers of this book. The author and their associates will not be held liable for any unintentional errors or omissions that may be found, and for damages arising from the use or misuse of the information presented in this book.

Readers should exercise their own judgment and discretion in interpreting and applying the information to their specific circumstances. This book is not intended to replace professional advice (especially medical advice, diagnosis, or treatment). Readers are encouraged to seek appropriate professional guidance for their individual needs.

The material in the book may include information by third parties. Third party materials comprise of opinions expressed by their owners. As such, the author of this book does not assume responsibility or liability for any third party material or opinions.

The publication of third party material does not constitute the author's guarantee of any information, products, services, or opinions contained within third party material. Use of third party material does not guarantee that your results will mirror our results. Publication of such third party material is simply a recommendation and expression of the author's own opinion of that material.

Whether because of the progression of the Internet, or the unforeseen changes in company policy and editorial submission guidelines, what is stated as fact at the time of this writing may become outdated or inapplicable later.

Wisdom University is committed to respecting copyright laws and intellectual property rights. We have taken reasonable measures to ensure that all quotes, diagrams, figures, images, tables, and other information used in this publication are either created by us, obtained with permission, or fall under fair use guidelines. However, if any copyright infringement has inadvertently occurred, please notify us promptly at wisdom-university@mail.net, providing sufficient details to identify the specific material in question. We will take immediate action to rectify the situation, which may include obtaining necessary permissions, making corrections, or removing the material in subsequent editions or reprints.

Made in the USA
Monee, IL
27 June 2024

a20a7c75-0601-4a86-8045-206ba2da3cb3R01